St. Louis Inland Type Foundry

Specimen Book and Catalog

A Price List of Printers' Supplies, Showing Types and Rules

St. Louis Inland Type Foundry

Specimen Book and Catalog
A Price List of Printers' Supplies, Showing Types and Rules

ISBN/EAN: 9783337166151

Printed in Europe, USA, Canada, Australia, Japan

Cover: Foto ©berggeist007 / pixelio.de

More available books at **www.hansebooks.com**

Specimen Book
and Catalog

❧

A Price List of Printers' Supplies, showing Types and
Rules in which are embodied all the Latest Ideas
that enable the Printer to produce Superior
Work in a most Economical Manner
Among which Betterments may
be especially mentioned the
Casting of Types on

Standard Line and Unit Sets

Inland Type Foundry
Saint Louis
February, 1897

INLAND TYPE FOUNDRY

▽▽▽

WM. A. SCHRAUBSTADTER,
President

OSWALD SCHRAUBSTADTER,
Vice-President

CARL SCHRAUBSTADTER, JR.,
Sec'y & Manager

The composition of this specimen book was done under the sup-
ervision of our Mr. N. J. Werner; the printing and binding
are the work of the Woodward and Tiernan Printing
Company; the paper was furnished by the Allen
Paper Company, of Saint Louis, and the ink by
the Fred'k H. Levey Company, New York.

Pertinent Facts

—————>#<—————

The Question of Quality............The keen competition which obtains in every branch of printing industry has produced the unfortunate result that the printer in purchasing material too often looks only at the **first cost** when placing his order. Unfortunate, because there is as great a difference in the various **qualities** of type as in the various grades of paper, ink, or printing itself, and **low priced** type is often by far the **dearest** in the end. While in the last decade great improvements have been made in the durability and accuracy of type, many of the foundries limit their competition to price alone, the result being an inferior article which is dear at any price. A few dollars saved by purchasing type which is not of the very best often involves a loss of many hundred times the amount. A large number of printers will testify that in accuracy, finish and improved wearing qualities alone, our products so far surpass all others as to justify a large increase in price, if we were disposed to make it. With the single exception of point bodies, until the Inland Type Foundry entered the field, little attention was given to the production of type which would economize the time of the compositor. The introduction of our new system of STANDARD LINE TYPE, with its many labor-saving features, is of vast importance to all printers, because by its use **superior** work, at a great **saving** of **time** and **labor,** is accomplished, as well as the **saving** of a large amount of **material,** and consequently **first cost.**

Type and Wages............In all offices, large and small, the amount invested in type is little when compared with the yearly expenditure for wages in the composing room. It therefore follows that type which accomplishes the saving of, say only 10 per cent in labor will cost the proprietor *nothing*, as it will pay for itself in a year, and thereafter yield a large profit on the investment. The printers who have adopted our system will testify that the **saving** in **composition** is large; in some cases being fully **50 per cent.** Not a single objection can be urged against the system. Being cast on point bodies, our type will work to advantage with that of other foundries. *It is unnecessary for an office to be entirely equipped with our material to derive benefit from* **Standard Line Type.** However, the larger the amount of our type, the greater the saving. We are convinced that every practical printer who once gives it a trial will find that it will pay him to replace his old type with STANDARD LINE, as the saving of labor in his composition will pay for it in a very short time.

Great Availability............Before going further it would be well to remark that it takes no great amount of study or experience to derive profit from any of our improvements. Being made to an absolute standard, and in a common-sense manner, even the youngest apprentice can avail himself of the benefits of our system, and improve both the quantity and quality of his work. It can never prove a hindrance, as have so many of the other schemes laid before the printer. At the same time, an experienced compositor can derive much greater benefits, since it works in with his experience, giving him, in fact, what he has been demanding for years. It enables him to do better work, while doing it faster and deriving more satisfaction therefrom.

STANDARD LINE............Perhaps the most important of our many improvements is STANDARD LINE.   All faces on each body, Romans, *Italics*, **Gothics, Antiques, 𝔊𝔢𝔯𝔪𝔞𝔫𝔰, 𝔗𝔢𝔯𝔱𝔰,** and all other job faces, line with one another at the bottom.   The following lines show a number of our 12-Point faces:

H H HHHHHHHH HⱧⱧH H HHⱧH H HHHH *H Ӿ ⱨⱨ 𝔥𝔟*

Great improvement *in* lining faces.

This enables the compositor to give **emphasis** to a word by using some heavy job face in the body of an article, or as in the following line, which shows a few words of our Gothic No. 1 and Old Style No. 9:

Perfect **lining** is secured by **our system**

With ordinary type the work would look like this:

Perfect **lining** is secured by **our system**

or it would be necessary to line the type by justifying with cardboard, consuming a great amount of time, precluding the possibility of setting it solid, or producing only an indifferent effect.

Italic Faces............It often happens that the larger offices having a number of Romans have no Italics for some of them.   As the Romans always differ in line, this requires a purchase of additional material.   With our type this is unnecessary, as *any* of our Italics will *line* correctly with *all* our Romans as well as with the job faces.

STANDARD TITLE LINE............In some cases—such as "two-line" fonts having no lower case—where casting faces on regular Standard Line would mean an unnecessary shoulder at the bottom, we cast them on Standard Title Line.   This uniform line for each body corresponds with the Standard Line of the next larger body.   Thus, any 10-Point Title Gothic will line with any 12-Point Gothic, or, for that matter, with any other 12-Point face, if a 2-Point lead be placed at the bottom of the 10-Point face.

STANDARD SCRIPT LINE ....In some other cases—such as Scripts, which have descenders of unusual length, which would kern too much if cast on Standard Line—we cast faces on Standard Script Line, which corresponds with the Standard Line of the next smaller body.   Thus, our 24-Point Commercial Script will line with any of our regular 20-Point job faces, if two 2-Point leads are placed below the latter, and with any of our 18-Point faces, if you place two 2-Point leads below and one 2-Point lead above the 18-Point.

Every Condition Considered............The Standard Title Line and Standard Script Line meet every contingency that the regular Standard Line could not meet. They have all the advantages possessed by Standard Line, and all three lines work harmoniously together.

German Faces........ Our system also enables the compositor to use Roman and German together, as on pages 137 to 139. In this case he needs but one set of figures, and the effect so frequently seen in job offices working in both these languages—1838564724561327—cannot possibly occur.

Leaders...........Our STANDARD LINE LEADERS will line with any of our Roman or job letters (as illustrated after the side-heads of this preface). This means not only a great saving of time, but of first cost in material and lessens the chances of mistakes and poor work.

Position of Line...........The line is placed in such a position that the different sizes of type faces will line with one another, if regular leads, slugs or quad lines are used to justify the difference between the bodies, as for instance the following lines:

BEST GRAND STANDARD LINE SYSTEM OF LINING TYPE

This saves a great deal of labor where it is desired to use different sizes of a series in combination as caps and small caps. Not only will all sizes of the same series line, but also those of two or more series of different faces; a condensed or fancy capital can therefore be lined with wider letters.

Use of Rules...........Furthermore, the line is placed in such a position that it will justify with ordinary 2-Point dotted..................or single_____ rule, justifying easily by the use of regular leads and slugs, as in the following example, in which the slugs and leads used for justification are shown black as though type high:

New York,████████████189█

Saint Louis, ████████████ *189*..........

This is of great importance in blank work and headings, as experienced printers can readily understand.

The Subject of Width...........Another very important improvement is our system of casting in width. Heretofore all type has in this respect been made totally without rule or method, not only making justification slow and laborious, but adding to the printers' annoyance by giving him sorts which do not agree with the original fonts. All our type is cast in width to agree with hardened steel standards, and is as accurate in this dimension as in body. Sorts *must* correspond with the rest of the font. Our unit of width is $\frac{1}{8}$-Point, but most of our faces are fitted to multiples of $\frac{1}{4}$-Point, and quite a number to multiples of $\frac{1}{2}$-Point and 1-Point. This reduces the number of widths of characters, makes justification and correction easier and in every way facilitates work. All our points and figures are cast on multiples of $\frac{1}{2}$-Point, and no special justifiers are required for tabular work, as the regular spaces can be used. Suppose, for instance, a customer wants to have inserted in the body of a price list a line of display type, you do not have to explain to him the difficulties in the way of justification, etc., incident to the old systems; with our type you can easily meet his wishes.

Year	Popula-tion	Increase	Increase Per cent.	Bonds
1860	2000		50	$40,000
1870	3000	1500	100	$30,000
1880	6000	3000	150	$25,000

The above specimen shows an ordinary statistical table, in which 8-Point Roman No. 20 (5-Point set), 10-Point Kelmscott and 12-Point Condensed Gothic No. 1 figures are used, justifying perfectly in every case by the use of regular spaces.

Width of Figures...........We have paid particular attention to our figures. As has before been mentioned, both in job faces and Romans, these all justify to multiples of one point, except en set figures in 5, 5½, 7, 9 and 11-Point bodies, and when other than en set, need no special justifiers, as the regular spaces can be used with them in tabular work. Except in some faces where the figure 1 is made narrower, all figures are uniform in width. Even in these few cases this character is of such thickness that the addition of a thin or other space on each side will make it justify with the other figures. See pages showing Point-Set Figures.

Old Style Figures...........Throughout the Old Styles the figures above and below the line have been abandoned, and while the Old Style design has been retained, these characters are made uniform in size and line, adapting them to all kinds of work and making them more legible and beautiful.

Small Caps and Nicks ....All small cap sorts likely to be confused with similar lower case letters have an extra nick, and all Old Style body letters have one more nick than the corresponding size of Modern Roman.

Kerning Letters...........The f's and j's in all cases, except Italics and Scripts, are made non-kerning, not only relieving the printer of the useless expense of the f combinations, but exempting him from danger of type breaking, and insuring better electrotypes. Except in a very few larger sizes, none of the descending letters in the lower case over hang the body.

Hight-to-Paper...........On account of our improved methods of making and fitting matrices, and superior casting machines, we are able to guarantee the perfectness of our type in respect to its hight, a point which every pressman will appreciate when he notes the **saving** in **make-ready** resulting from the use of our products.

Our Metal...........This is of a new composition. We *guarantee* our type to wear longer, print better and to give finer results in electrotyping and stereotyping than any other. In finish and accuracy it is unapproachable.

Our Foundry...........Our equipment is the very best possible. We have endeavored to profit by our former experience, and have established a plant of the latest and most improved machinery, which is mostly of our own design and manufacture.

Leads, Slugs and Brass Goods............Our Brass Rule department is complete in every detail, and in addition to the specimens shown herein we can furnish special designs. We solicit jobbing in leads, slugs and all brass goods. Having complete machine and wood-working shops, we can attend to press repairing, as well as designing of special machinery.

Other Makers' Goods............In addition to our own productions, we carry in stock a complete line of type made by the other foundries casting on Standard Line, thus giving the printer a large stock from which to make his selection. We also keep everything required by the printer in the way of cases, cabinets, stands and other wood goods, leads, slugs, furniture, inks, tapes, quoins, and all sorts of printers' supplies. We are agents for all makes of paper cutters, bookbinders' machinery, and printing presses, including the best makes of Old Style Gordons, cylinder presses, and such presses and other novelties as are made by Golding & Co. We buy and sell second-hand machinery, and from time to time issue lists, which may be had on application.

Our Relation to Trusts............We desire to emphasize the fact that ours is the only type foundry in the South and west of the Mississippi (excepting San Francisco) that is **not** connected with the Type Trust. Having no connection with any combination, we are in a position to make our prices as low as is consistent with the superiority of our goods. Terms and discounts sent on application.

In conclusion.......We cannot but felicitate ourselves upon the advances we have made since we issued our first specimen book, in February, 1895. The space then devoted to Standard Line Roman and job type was but 55 pages, compared with 144 pages in this edition showing an assortment extensive enough to enable any one to select a complete office, however large, and thus get the full benefit of Standard Line. At that time, when we were barely a year old, our innovations were decried as being impractical, and our plans denounced as visionary. The large business we have built up in spite of the keen competition and panicky times shows how quick printers are to appreciate a really good thing. Several other foundries have, with our instruction and assistance, adopted the Standard Line system. Some other foundries claim that their type possesses advantages similar to those of our **Standard Line,** but we warn printers not to believe such statements. To carry out the system properly requires a thorough understanding of all the conditions that **Standard Line** has met, and, from tests we have made, we are prepared to demonstrate that these thieving concerns are either ignorant of all of these conditions or altogether incompetent to meet them.

Correspondence............We solicit correspondence, and shall take pleasure in giving particulars about our manufactures and those for which we are agents. An illustrated catalog of machinery and printers' supplies is included at the end of this book. Circulars and prices of special goods will be sent on request.

INLAND TYPE FOUNDRY.

SAINT LOUIS, February, 1897.

READ CAREFULLY!

THE **Prices** in this book are subject to **discount.** If you have no discount sheet, send for one and paste it inside the cover of this book.

PRICES and discounts are subject to change without notice.

THIS book is issued to our patrons to aid them in making selections, and we would remind them that it is only necessary to give the **Size, Name** and **Number** (if any) of the Type, or Number of the Rule desired, to insure the accurate filling of their orders.

Do not cut this book under any circumstances.

TAKE a separate line or paragraph for each item in your order.

WHEN ordering a font of type, or sorts, remember that every foundry has its own **special numbers,** and that the number, name of type, and specimen book should always be given. Our "10-Point Roman No. 20," for instance, doesn't mean that the face is similar to one designated by the same number in any other foundry's specimen book.

WHEN you are ordering type for newspaper or book work, state whether or not you want the fonts to contain **Italics, Leaders, Fractions** or **Commercial Marks,** as none of these are sent unless **specially** ordered.

STATE particularly what styles of **Fractions** you want to accompany your Roman fonts. Make your own selections from the specimens shown, as we prefer not to select for you.

JOB fonts, such as Titles, Antiques, Ornamented, etc. (except Scripts), are put up without **Spaces and Quads,** and if you want them for any size of body it will be necessary for you to order them separately.

IN ordering fonts of job letter by **weight** specify whether or not spaces and quads are to be included, otherwise we shall consider that we are authorized to use our own judgment.

IN ordering **Sorts** to match type you already have, state the Name and Number of the Face, as well as the Size of the Body; or, if you cannot do this send a capital "H" and lower case "m" (as little used as possible) of the font the sorts are to work with. Also state the number of pounds or ounces you want of each particular Sort.

AN ordinary cap box holds about 6 ounces of type, the lower case "e" about 3 pounds, the "a" about 2 pounds, the "b" about 15 ounces, and the "k" box about 6 ounces, when full.

WE very often receive **samples** that are so much worn that it is an utter impossibility to determine the face. Send the very newest or least worn that you can find, and so prevent trouble, delay and dissatisfaction on both sides.

WHEN you write "Inclosed you will find samples," always give another look in the envelope to corroborate the assertion. We frequently look for the samples in vain. This omission causes the writing of two letters and the loss of some time in the filling of orders.

OUR Borders, Ornaments and Cuts are numbered systematically. The first figures of each number indicate the body. Thus, Cut No. 7201 is cast on 72-Point body, and Border No. 1801 on 18-Point body.

WHEN you want **Leads, Advertising Rules** or **Dashes,** send one of your leads; or, if you order them cut to certain ems long, be particular to state what particular ems you mean, whether 6-Point, 8-Point, 10-Point or 12-Point. Unless otherwise specified we shall understand 12-Point ems.

SHOULD you order Brass Rule to be mitered to any size, give either the "inside" or "outside" measurement, and if it is to be cut from Double Rules state whether you wish the heavy line inside or outside.

IN ordering **Printing Inks** or **Bronzes,** state the quantity and price per pound or ounce of each kind you want.

WHEN you want **Cbases,** give the "INSIDE" or "OUTSIDE" measurement, or, better still, send a diagram.

Shipping directions should not be neglected. State whether goods are to be sent by water or railway, freight or express, naming the route, otherwise we will assume that we are to use our own judgment. Drayage is charged on freight shipments.

THE weight of a single parcel by mail is limited to 4 pounds. Type lots weighing over 4 pounds can be divided as required to send by mail. The rate of postage is 1 cent per ounce. But it is safer and generally cheaper to forward over 2 pounds by express, and over 50 pounds by freight.

IF goods fail to arrive within a reasonable time, notify us and we will send a tracer after them. If freight charges seem too heavy, advise us of the rate and amount paid. On the arrival of goods check them by the bill, and notify us at once of the amount paid.

IMMEDIATELY upon type being received, if it be a job or display font, **take a proof** of it, and examine carefully to see if every letter is in the font, before laying in the case. We furnish no sorts upon any claim for shortage, unless a proof of the font as it appeared before being laid is sent with the claim. Every font is guaranteed complete.

THE regular paging galley used by founders, which is about 5x6 inches inside, will be found of great advantage in laying type.

WHEN you ship *Old Type,* see that your name is on each box, as well as our address. Notify us by mail of the shipment, and pack type, leads, electrotypes, and brass, in separate parcels. Do not melt them up, and do not include zinc etchings, bottle tops, or anything except printers' metal, otherwise the entire lot will be rejected.

REMIT by post-office or express money orders, or draft on Saint Louis, Chicago or New York. On personal checks our banks make a charge for collection, which will be deducted from your remittance. Small amounts may be sent in postage stamps.

WHEN small items are desired sent by mail, enough extra cash should be remitted to cover postage.

SHOULD you desire to open an account at a foundry where you have no acquaintance, send references or the money. Or, if you wish a cut or small package sent C. O. D., send enough money to pay charges at least, or you may suffer from delay while the founder writes you for a remittance, or for references as to your financial standing. Confidence is of especial value in rendering business relations pleasant, and nothing else so tends to give confidence as ready and prompt payments.

STANDARD LINE JUSTIFICATION TABLE, showing how to justify any two different bodies with one another so that their faces will line accurately together.

REGULAR STANDARD LINE FACES

	2	5	6	7	8	9	10	11	12	14	16	18	20	24	30	36	42	48	54	60	72
72 above	.57	.54	.53	.53	.52	.51	.50	.50	.49	.47	.45	.44	.42	.39	.35	.30	.24	.18	.12	. 6.	72
below	13	13	13	12	12	12	12	11	11	11	11	10	.10.	9.	7.	6.	6.	6.	6.	6.	
60 above	.51	.48	.47	.47	.46	.45	.44	.44	.43	.41	.39	.38	.36	33	29	24	18	12	. 6.	60	
below	7	7	7	6	6	6	6	5	5	5	5	4	4.	3.	1.	0.	0.	0.	0.	6.	
54 above	.45	.42	.41	.41	.40	.39	.38	.38	.37	.35	.33	.32	.30	27	23	18	12.	. 6.	54		
below	7	7	7	6	6	6	6	5	5	5	5	4	4.	3.	1.	0.	0.	0.	6.		
48 above	.39	.36	.35	.35	34	33	32	32	31	29	27	26	24	21	17	12	. 6.	48			
below	7	7	7	6	6	6	6	5	5	5	5	4	4.	3.	1.	0.	0.	6.			
42 above	.33	.30	29	29	28	27	26	26	25	23	21	20	18	15	11.	. 6	42				
below	7	7	7	6	6	6	6	5	5	5	5	4	4.	3.	1.	0.	6.				
36 above	.27	24	23	23	22	21	20	20	19	17	15	14	12	. 9.	. 5.	36					
below	7	7	7	6	6	6	6	5	5	5	5	4	4.	3.	1.	6.					
30 above	22	19	18	18	17	16	15	15	14	12	10.	9.	7.	4.	30						
below	6	6	6	5	5	5	5	4	4	4	4	3	3.	2.	6.						
24 above	18.	15.	14	13	12	11	11	10.	8.	6.	5.	3.	24								
below	4	4	4	3	3	3	3	2	2	2	1	1.	6.								
20 above	15	12	11	11	10.	9.	8.	8.	7.	5.	3.	2.	20								
below	3	3	3	2	2	2	2	1	1	1	1.	0.	6.								
18 above	13.	10.	9.	9.	8.	7.	6	6.	5.	3.	1.	18.									
below	3	3	3	2	2	2	2	1	1	1	1.	6.									
16 above	12.	9.	8.	8.	7.	6.	5.	5.	4.	2	16.										
below	2	2	2	1	1	1	1	0	0	0.	6.										
14 above	10.	7.	6.	6.	5.	4.	3.	3.	2.	14.											
below	2	2	2	1	1	1	1	0	0.	7.											
12 above	8.	5.	4.	4.	3.	2.	1.	1.	12.												
below	2	2	2	1	1	1	1	0.	7.												
11 above	7.	4.	3.	3.	2.	1	0	11.													
below	2	2	2	1	1	1	1.	0.													
10 above	7.	4.	3.	3.	2.	1	10														
below	1	1	1	0	0.	0.	0.														
9 above	6	3	2	2	1	9															
below	1	1	1	0	0.	0.															
8 above	5.	2	1	1	8																
below	1	1	1	0.	0.																
7 above	4.	1	0	7																	
below	1	1	1.	0																	
6 above	4.	1	6.																		
below	0.	0.	0.																		
5 above	3.	5																			
below	0.	0.																			
2 above	2.	3.	4.	5.	6.	7.	8.	9	10.	12	13.	15.	17	21	27	33	39	45	51	57	63
below	0.	0.	0.	0.	0.	0.	0.	0.	1.	1.	1.	1.	1.	1.	1.	1.	1.	1.	1.	1.	7

STANDARD TITLE LINE FACES

Explanation of the Table.—Each group of figures states the amount in points, either of leads, slugs or quad lines, to be placed **above and below** the smaller body, when justifying any two bodies together in order to make their faces match on the same line.

The proper justification-figures for the combination of any two bodies are ascertained by noting the group found where the horizonal line of groups of one body meets the vertical line of groups of the other body.

Thus, for instance, to combine a 30-Point face with a 24-Point face, as caps and small caps, take the group of figures found where the 30-Point line crosses the 24-Point line. For STANDARD LINE faces the group reads "4 above, 2 below," indicating the use of two 2-Point leads above and one 2-Point lead below the 24-Point body, to justify its face in line with that of the 30-Point. For **Standard Title Line** faces the group reads "6 above, 0 below."

The figures **above** the diagonal line of bodies are given for faces cast on our regular **Standard Line**, and those **below** this line are for all faces cast on **Standard Title Line.**

The groups following the figure **2** are for the justification of **2-Point Rule,** either single or dotted, with every body, insuring accurate lining of such in blank work or date-lines.

PRICES OF STANDARD LINE TYPE

— ·- ⚏ —

THE FIRST CLASS comprises Romans—Modern and Old Style (Including French Old Style), with proportionate Italic, and Germans, in fonts of 25 pounds and over; also Leaders and Spaces and Quads. The SECOND CLASS comprises Romans, Italics and Germans In fonts under 25 pounds; also all standard job faces, such as Gothics, Antiques, Full-Faces, Two-Lines, etc., except Extra Condensed. The THIRD CLASS comprises Scripts, Hair-Lines, Extra Condensed and all Shaded, Ornamental and Patented Faces, Accents, Signs, etc.

Sold at following prices per pound:

BODY	FIRST CLASS 25 lbs. and over	1000 lbs. and over	*Poster Fonts 25 lbs.	*Poster Fonts 50 lbs	SECOND CLASS	THIRD CLASS
3-Point						$3.60
3½-Point						3.60
4-Point						3.60
4½-Point	$1.62					3.20
5-Point	1.20		$1.60		$2.00	2.80
5½-Point	.74	$0.73	1.30		1.60	2.40
6-Point	.64	.63	1.00		1.28	2.00
7-Point	.56	.55	.90		1.12	1.80
8-Point	.53	.52	.80		1.00	1.60
9-Point	.50	.49	.70		.90	1.44
10-Point	.48	.47	.65		.82	1.30
11-Point	.46	.45	.60		.78	1.22
12-Point	.45	.44	.54		.74	1.16
14-Point	.45		.52		.70	1.12
16-Point	.45		.52		.66	1.00
18-Point	.45		.52		.66	1.00
20-Point	.45		.52		.66	.94
24-Point	.45		.52		.64	.90
30-Point	.45		.50		.64	.86
36-Point	.45			$0.50	.62	.82
42-Point	.45			.50	.60	.78
48-Point	.43			.50	.60	.72
54-Point	.43			.50	.60	.72
60-Point	.43			.50	.60	.72
72-Point	.43			.50	.60	.72

*Poster fonts Include spaces and quads.

ACCENTS, SIGNS, ETC.
CAST TO ORDER—Prices per pound.

4½-Point $3.20	10-Point $1.30	30-Point $0.86
5-Point.............. 2.80	11-Point................ 1.22	36-Point .... .82
5½-Point.......... 2.40	12-Point................ 1.16	42-Point .... .78
6-Point.............. 2.00	14-Point................ 1.12	48-Point .... .72
7-Point.............. 1.80	16-Point................ 1.00	54-Point .... .72
8-Point.............. 1.60	18-Point................ 1.00	60-Point .... .72
9-Point.............. 1.44	20-Point................ .94	72-Point .... .72
	24-Point.............., .90	

SUPERIORS AND INFERIORS
Prices per pound.

5-Point $2.80	7-Point...................$1.80	11-Point....................$1.22
5½-Point 2.40	8-Point................... 1.60	12-Point.................... 1.16
6-Point 2.00	9-Point................... 1.44	14-Point.................... 1.12
	10-Point................... 1.30	

PIECE FRACTIONS
Prices per pound.

8-Point...............................$3.60	11-Point................................... $2.40
9-Point............................... 3.20	12-Point................................... 2.00
10-Point............................... 2.80	14-Point................................... 1.44

We furnish no quantity for less than 25 cents net.

Special Accents and Marked Letters will be cut to order, the cost being from $2.00 to $3.00 for making each matrix.

HOW TO ESTIMATE

To ascertain the Quantity of Plain Type required for a newspaper, magazine, or other work find the number of square inches in the matter, and divide the same by four; the quotient will be the approximate weight. As it is impossible to set the cases entirely clear, it is necessary to add 25 per cent. to large fonts, and 33 per cent. to small, to allow for dead matter.

Leaded Matter requires about 25 per cent. less type than Solid Matter. The following table shows the weight of 2-Point Leads required to lead 1000 ems Solid, and contained in 1000 ems Leaded matter.

TABLE OF WEIGHT OF 2-POINT LEADS IN MATTER

Per 1000 ems	To Lead Solid Matter	In Leaded Matter	Per 1000 ems	To Lead Solid Matter	In Leaded Matter
5-Point	7½ oz.	5½ oz.	9-Point	13½ oz.	11 oz.
5½-Point	8½ oz.	6 oz.	10-Point	15½ oz.	12½ oz.
6-Point	9½ oz.	7½ oz.	11-Point	16½ oz.	14 oz.
7-Point	11½ oz.	9 oz.	12-Point	19 oz.	16½ oz.
8-Point	13 oz.	10½ oz.			

TABLE OF MEASUREMENT FOR NEWSPAPER ESTIMATES

WIDTH OF STANDARD COLUMN: 13 EMS PICA.	5½-Pt.	6-Point	7-Point	8-Point	9-Point	10-Point
No. of Ems per Line	28½	26	22½	19½	17½	16
No. of Lines in 1000 Ems	35½	38⅓	44½	51½	57½	62½
No. of Inches in 1000 Ems	2¾	3¼	4⅜	5⅝	7⅛	8⅝
No. Ems in Column, 4-Col. Folio	5,040	4,325	3,175	2,465	1,950	1,610
No. Ems in Column, 5-Col. Folio	6,505	5,615	4,115	3,200	2,525	2,085
No. Ems in Column, 6-Col. Folio	7,180	6,160	4,515	3,510	2,770	2,290
No. Ems in Column, 7-Col. Folio	7,900	6,985	4,970	3,685	3,050	2,520
No. Ems in Column, 8-Col. Folio	8,630	7,410	5,440	4,220	3,330	2,775
No. Ems in Column, 9-Col. Folio	9,310	8,030	5,885	4,575	3,615	2,970

TABLE OF MEASUREMENT FOR BOOK TYPE ESTIMATES

MEASURE, 25 EMS PICA.	6-Point	8-Point	9-Point	10-Point	11-Point	12-Point
No. 2-Point Leads to Pound	31	31	31	31	31	31
No. 3-Point Leads to Pound	21	21	21	21	21	21
No. Ems in Line	50	38	23⅓	30	27½	25
No. Lines to 1000 Ems	20	26½	30	33½	36½	40
No. of Inches to 1000 Ems	1¾	3	3¾	4⅝	5⅝	6⅝
No. 2-Point Leads to lead 1000 Ems	13	21	25	27	32	33
No. 3-Point Leads to lead 1000 Ems	12	19	23	25	28	31

TABLE OF STANDARD SIZES OF NEWSPAPERS

The following are the regular sizes of newspapers adopted by the auxiliary printers. As a matter of convenience in the buying of printing material and paper, we would advise parties planning new newspapers to adopt one of these sizes. The width of column is 13 ems 12-Pt.

	Size of Paper	Column Rule	Head Rule
5-Column Folio	20x26 inches	17¾ inches	11⅛ inches
6-Column Folio	22x31 inches	19¾ inches	13⅜ inches
7-Column Folio	24x35 inches	21¾ inches	15⅝ inches
8-Column Folio	26x40 inches	23¾ inches	17⅞ inches
9-Column Folio	28x44 inches	26 inches	20 inches
4-Column Quarto	22x31 inches	13¾ inches	8⅞ inches
5-Column Quarto	26x40 inches	17¾ inches	11⅛ inches
6-Column Quarto	30x44 inches	19¾ inches	13⅜ inches
7-Column Quarto	35x48 inches	21¾ inches	15⅝ inches

ESTIMATE FOR JOB OFFICE

In Connection with a Seven or Eight-Column Newspaper

1 Challenge-Gordon Job Press 10x15 inches inside of chase	$250.00
1 Marble Imposing Stone, 24x36, with Stand	11.00
1 dozen Steel Quoins and Key	3.00
1 Stained Cabinet, 16 Two-thirds Job Cases, Galley Top	21.00
1 Job Stand, with Racks for 12 Full and 12 Two-thirds Cases	5.50
2 pairs News Cases	1.60 3.20
1 Labor-Saving Rule Case	1.15
1 Labor-Saving Lead and Slug Case	1.00
10 Job and Triple Cases	.90 9.00
3 pairs News Cases	1.60 4.80
10 Two-thirds Job Cases	.75 7.50
1 Job Galley, 10x16	3.00
1 Wood Composing Stick, 24-inch	1.20
1 Composing Stick, 14-inch	1.60
20 pounds 2-Point Leads, long	.16 3.20
50 pounds Labor-Saving 2-Point Leads and 6-Point Slugs	.25 12.50
25 pounds Labor-Saving Metal Furniture	.25 6.25
2 pounds Labor-Saving Single Rule, 2-Point	1.75 3.50
2 pounds Labor-Saving Dotted Rule, 2-Point	1.75 3.50
2 pounds Labor-Saving Double Rule, 6-Point	1.50 3.00
5 pounds 8-Point Fine-dot Leaders (Standard Line), for blanks	.53 2.65
1 22½-inch Paper Cutter	80.00
1 Little Giant Lead and Rule Cutter	8.00
5 pounds Job Black Ink	.50· 2.50
1 pound Fine Job Black	2.00
Colored Inks, say	6.00
25 pounds 6-Point Roman, Old Style or Modern (Standard Line)	.64 16.00
25 pounds 10-Point Roman, Old Style or Modern (Standard Line)	.48 12.00
25 pounds 18-Point Woodward, Poster font	.52 13.00
20 fonts Woodward, Inland and Gothics, say	55.00
12 fonts Cosmopolitan and Scripts, say	65.00
12 fonts Saint John, Kelmscott, Caledonian Italics, etc., say	50.00
3 fonts Wood Type, about	30.00
1 half-case Labor-Saving Wood Furniture, with Case	5.00
30 yards 6-Point and 12-Point Reglet	.60
1 dozen Gage Pins	.60
Cuts, Borders, Ornaments and Dashes, say	18.00
Spaces and Quads for Job Type	14.00
Subject to Discount.	$735.25

ESTIMATES FOR WEEKLY NEWSPAPERS

In these estimates, 10-Point and 8-Point are inserted merely to show the quantity of type required. Other sizes may be substituted, varying the expense but little.

ESTIMATE FOR 6-COLUMN PAPER WITH ARMY PRESS

Army Press, 14x20 inches	$ 60.00
3 Brass-Lined Galleys, 3½x23⅝ inches	2.00 6.00
2 Six-inch Composing Sticks	.75 1.50
3 pairs News Cases	1.60 4.80
6 Job and Triple Cases	.90 5.40
1 News Stand	3.75
Mallet, 30c.; Planer, 40c.; Lye Brush, 40c.	1.10
10 pounds News Ink	.20 2.00
10 Column Rules (6-Point), 5 short	.45 4.50
3 Head Rules (6-Point), 2 Double, 1 Parallel	.40 1.20
20 Advertising Rules, 4c.; 10 Double, 6c., and 10 Single Dashes, 6c.	2.00
20 pounds Leads and Slugs, cut to measure	.18 3.60
Head to Paper	2.50
75 pounds 10-Point Roman (Standard Line)	.48 36.00
50 pounds 8-Point Roman (Standard Line)	.53 26.50
1 font 8-Point Woodward	2.25
1 font 8-Point Extended Woodward	2.25
1 font 10-Point Woodward	2.50
1 font 10-Point Extended Woodward	2.50
1 font 10-Point Condensed Woodward	2.50
1 font 18-Point Woodward, $3.20; 1 pound 18-Point Spaces and Quads, 45c.	3.65
Subject to Discount.	$176.50

ESTIMATE FOR 7-COLUMN PAPER WITH ARMY PRESS

Army Press, 16⅝x22¾ inches		$ 85.00
4 Brass-Lined Galleys, 3½x23⅛ inches	2.00	8.00
3 Six-inch Composing Sticks	.75	2.25
3 pairs News Cases	1.60	4.80
7 Job and Triple Cases	.90	6.30
1 News Stand, double		3.75
Mallet, 30c.; Planer, 40c.; Lye Brush, 40c.		1.10
10 pounds News Ink	.20	2.00
12 Column Rules (6-Point), 6 short	.50	6.00
3 Head Rules (6-Point), 2 Double, 1 Parallel	.45	1.35
25 Advertising Rules, 4c.; 12 Double, 6c., and 12 Single Dashes, 6c.		2.44
30 pounds Leads and Slugs, cut to measure	.18	5.40
Head to Paper		2.75
100 pounds 10-Point Roman (Standard Line)	.48	48.00
50 pounds 8-Point Roman (Standard Line)	.53	26.50
1 font 8-Point Woodward		2.25
1 font 8-Point Extended Woodward		2.25
1 font 10-Point Woodward		2.50
1 font 10-Point Condensed Woodward		2.50
1 font 10-Point Extended Woodward		2.50
1 font 12-Point Woodward		2.80
1 font 12-Point Condensed Woodward		2.80
1 font 18-Point Woodward		3.20
1 font 18-Point Condensed Woodward		3.20
2 pounds each 12-Point and 18-Point Spaces and Quads	.45	1.80

Subject to Discount. $231.44

ESTIMATE FOR 6-COLUMN FOLIO—Size, 22x31 Inches

The right hand column of figures gives the amount when it is intended to use auxiliary insides.

No. 3 Washington Press	$200.00			$200.00
1 pair Chases		10.00		10.00
1 Sixteen-inch Roller Frame, Core and Casting		4.10		4.10
1 Set Iron Side and Foot Sticks		3.35		3.35
4 Single Patent-Lined Galleys	2.00	8.00	2	4.00
3 Six-inch Composing Sticks	.75	2.25	3	1.50
2 Job Stands, with Racks	5.00	10.00		10.00
5 pairs News Cases	1.60	8.00	4	6.40
10 Job Cases	.90	9.00	8	7.20
30 yards Reglet and Furniture, assorted	.05	1.50	20	1.00
1 Cast-Steel Shooting Stick, medium, 75c.; 1 Planer, 40c.		1.15		1.15
100 Boxwood Quoins, 60c.; 1 Hickory Mallet, large, 40c.		1.00		1.00
1 Lye Brush, Tampico		.30		.30
50 Advertising Rules, 13 ems 12-Point	.04	2.00	30	1.20
20 each Double and Parallel Dash Rules	.08	3.20	10 each	1.60
20 Single Dash Rules	.06	1.20	10	.60
12 each Double and Parallel Cross Rules	.06	1.44	10 each	1.20
30 pounds Leads and Slugs	.18	5.40	20	3.60
24 Metal Foot Slugs	.04	.96	12	.48
20 Column Rules (6-Point) 19¾ inches long	.45	9.00	10	4.50
4 Double Head Rules (6-Point) 13¾ in. long	.40	1.60	2	.80
150 pounds 10-Point Roman (Standard Line)	.48	72.00	125	60.00
100 pounds 8-Point Roman (Standard Line)	.53	53.00	75	39.75
5 pounds 10-Point Italic (Standard Line)	.48	2.40		2.40
15 pounds 8-Point Leaders, Fractions, etc.	.53	7.95	10	5.30
2 fonts 8-Point Woodward	2.25	4.50	1	2.25
2 fonts 10-Point Woodward	2.50	5.00	1	2.50
1 font 12-Point Woodward		2.80		2.80
1 font 18-Point Woodward		3.20		3.20
1 font 8-Point Extended Woodward		2.25		2.25
1 font 10-Point Extended Woodward		2.50		2.50
1 font 10-Point Condensed Woodward		2.50		2.50
1 font 12-Point Condensed Woodward		2.80		2.80
1 font 18-Point Condensed Woodward		3.20		3.20
2 pounds each 12-Point and 18-Point Spaces and Quads	.45	1.80		1.80
Electrotype Head		2.75		2.75
Borders and Ornaments		2.00		2.00
1 Imposing Stone, 30x60		12.00		12.00

Subject to Discount. $466.10 $413.98

☛In ordering, always state what Italic, Leaders, Commercial Marks, Fractions, etc., are required, as none are sent unless specially ordered.

The right hand column of figures gives the amount when it is intended to use auxiliary insides

No. 4 Washington Hand Press		$225.00		$225.00
7-column News Chases, in halves		11.00		11.00
1 Eighteen-Inch Roller Frame, Core and Casting		4.55		4.55
1 set Iron Side and Foot Sticks		3.75		3.75
6 Single Patent-Lined Galleys	2.00	12.00	4	8.00
4 Six-Inch Composing Sticks	.75	3.00	3	2.25
4 Job Stands, with Racks	5.00	20.00	2	10.00
6 pairs News Cases	1.60	9.60	5	8.00
10 Job Cases	.90	9.00	8	7.20
50 yards Reglet and Furniture, assorted		2.50	30	1.50
Mallet, Planer, Shooting Stick and 100 Quoins		2.15		2.15
1 Lye Brush, 40c; 1 Saw, $1.25; 1 Wood Miter Box, 40c		2.05		2.05
75 Advertising Rules	.04	3.00	60	2.40
25 each Double and Parallel Dash Rules	.08	4.00	16 each	2.56
25 Single Dash Rules	.06	1.50	18	1.08
16 each Double and Parallel Cross Rules	.06	1.92	10 each	1.20
30 pounds Leads and Slugs, cut	.18	5.40	20	3.60
28 Metal Foot Slugs (18-Point)	.04	1.12	14	.56
24 Column Rules (6-Point) 21¾ inches long	.50	12.00	12	6.00
5 Double Head Rules (6-Point) 15⅝ in. long	.45	2.25	3	1.15
2 Parallel Head Rules (5-Point) 15⅝ in. long	.40	.80		.80
200 pounds 10-Point Roman (Standard Line)	.48	96.00	150	72.00
150 pounds 8-Point Roman (Standard Line)	.53	79.50	100	53.00
7 pounds 8-Point Leaders and Fractions	.53	3.71		3.71
4 fonts 8-Point Woodward	2.25	9.00	3	6.75
2 fonts 8-Point Extended Woodward	2.25	4.50		4.50
3 fonts 10-Point Woodward	2.50	7.50	2	5.00
2 fonts 10-Point Extended Woodward	2.50	5.00	1	2.50
1 font 12-Point Woodward		2.80		2.80
1 font 18-Point Woodward		3.20		3.20
Other Display Type, and Spaces and Quads, say		25.00		23.00
Heading for Paper, say		3.00		3.00
Inland Borders and Ornaments, say		5.00		4.00
1 Marble Imposing Stone, 36x72		18.00		18.00
Subject to Discount.		$598.80		$506.26

No. 5 Washington Hand Press		$250.00		$250.00
8-column News Chases, in halves		13.00		13.00
1 Twenty-inch Roller Frame, Core and Casting		5.00		5.00
1 set Iron Side and Foot Sticks		4.20		4.20
6 Single Patent-Lined Galleys	2.00	12.00	4	8.00
4 Job Stands, with Racks	5.00	20.00	2	10.00
4 Six-Inch Composing Sticks	.75	3.00	3	2.25
8 pairs News Cases	1.60	12.80	6	9.60
12 Job Cases	.90	10.80	10	9.00
50 yards Reglet and Furniture, assorted		2.50	30	1.50
Mallet, Planer, Shooting Stick and 100 Quoins		2.15		2.15
1 Saw, $1.25; 1 Lye Brush, 45c; 1 Wood Miter Box, 40c		2.10		2.10
100 Advertising Rules	.04	4.00	75	3.00
75 Fancy Brass Dashes	.10	7.50	50	5.00
16 each Double and Parallel Cross Rules	.06	1.92	10 each	1.20
35 pounds News Leads	.18	6.30	20	3.60
32 Metal Foot Slugs	.04	1.28	16	.64
28 Column Rules (6-Point) 23¾ inches long	.55	15.40	14	7.70
5 Double Head Rules (6-Point) 17⅞ in. long	.50	2.50	3	1.50
2 Parallel Head Rules (5-Point) 17⅞ in. long	.45	.90		.90
250 pounds 10-Point Roman (Standard Line)	.48	120.00	150	72.00
175 pounds 8-Point Roman (Standard Line)	.53	92.75	125	66.25
7 pounds 8-Point Leaders and Fractions	.53	3.71		3.71
4 fonts 8-Point Woodward	2.25	9.00	3	6.75
4 fonts 10-Point Woodward	2.50	10.00	3	7.50
2 fonts 12-Point Woodward	2.80	5.60		5.60
2 fonts 18-Point Woodward	3.20	6.40		6.40
Other Display Type, and Spaces and Quads, say		26.00		22.00
Heading for Paper, say		3.25		3.25
Inland Borders and Ornaments, say		5.00		4.75
1 Marble Imposing Stone, 36x72		18.00		18.00
Subject to Discount.		$677.06		$556.55

Five-column Quarto outfit amounts to about the same as an 8-column Folio.

IMPORTANT NOVELTIES!

IMPROVED DASHES AND DOUBLE QUOTATION MARKS

TO CARRY out suggestions made to us by a number of our most progressive customers, who were dissatisfied with the ordinary Em Dash (when used as a mark of punctuation) and the Single Commas and Apostrophes used for quoting, we have cut a new series each of Improved Em Dashes and Double Quotation Marks, and take pleasure in presenting specimens of them in the following matter. We have no doubt that they will be generally welcomed.

The Dashes are somewhat heavier in face than the old Dashes, and are cut slightly shorter in length, so that—while they are cast on em bodies—it is no longer necessary to set thin or hair spaces between them and adjoining words, a feature which every compositor will appreciate. They are provided with an extra nick, to more readily distinguish them from ordinary Dashes.

"Improved Em Dashes" are supplied in proportionate quantity with all our Modern and Old Style Romans fonts, without extra charge, the quantity of ordinary Dashes in the fonts being reduced to that extent. These Dashes are also put up separately in 1-pound fonts, which are furnished at the same prices as the Romans. They will work with any face, whether this be of our make or not, and printers will find this a convenient way of procuring them.

The "Double Quotation Marks" comprise those for beginning and ending a quotation. Those for the beginning differ from the turned Commas, having the "tails" descending instead of ascending, and matching in reverse those for ending—this being the form most approved by tasteful book printers. To obviate the necessity of setting thin spaces after them, the "Quotes" for beginning are cast with the appropriate space after each.

"Double Quotation Marks" are put up in 1-pound fonts containing both characters, and are supplied at Roman prices. They are not furnished with the regular Romans, unless specially ordered. Printers will do well to order one or more of these small fonts, since they come in this convenient shape.

12-POINT
Will. Eskew, Wellston, Ohio—"The more I learn of your system the better I like it."

11-POINT
Albert W. Dennis, Lynn, Mass.—"I have often wondered why some one did not attempt it before."

10-POINT
D. B. Landis, Lancaster, Pa.—"Your Standard Line idea is something printers should have had years ago."

9-POINT
Charles T. Henderson, Toulon, Ill.—"I am pleased with your faces, and am also greatly taken with Standard Line."

8-POINT
W. H. Bevis, Pawtucket, R. I.—"Your system of type-making is certainly a great improvement over the common way, and I have no doubt the type will pay for itself in the time saved."

7-POINT
Henry Hahn, foreman "Northwestern Miller," Minneapolis, Minn.—"We are eagerly watching your output and shall try to give you our business, putting in your type as fast as the old wears out."

6-POINT
E. L. Wepf, Denver, Colo.—"I like your type not only as to its lining feature, but the making of figures with Old Style type and above the line of the same height. When I need more type you will hear from me'"

5½-POINT
W E. Fleming, Belleville, Ill.—"I am highly elated over your Standard Line system, and wonder why founders who have devoted a life-time to the business have never catered to the printers' convenience in casting type."

5-POINT
E. D. Wescott, Reading, Pa.—"You have made another step towards perfection. The point system was a long one, you have made another. I think the tests have all been made, and when I lay your type I will merely be putting a good thing into practical use."

STANDARD LINE LEADERS

»»›‹‹«‹

Our **Standard Line Leaders** are cast from a new, hard and tough composition. We guarantee that the **dots** will **not** break off and that they **wear longer** than any others. They are made in four styles, as shown below, and are supplied in 2-pound and 5-pound fonts, at the following prices per pound:

5-Point	$1.20	10-Point	$0.48
5½-Point	.74	11-Point	.46
6-Point	.64	12-Point	.45
7-Point	.56	14-Point	.45
8-Point	.53	16-Point	.45
9-Point	.50	18-Point	.45

»»›‹‹«‹

Standard Line Round-Dot Leaders No. 1, (two dots to em), cast on all bodies from 5-Point to 18-Point, in en, em, 1½-em, 2-em and 3-em widths.

... .

Standard Line Fine-Dot Leaders No. 2, cast on all bodies from 6-Point to 14-Point, in en, em, 1½-em, 2-em and 3-em widths.

Standard Line Round-Dot Leaders No. 3, (one dot to em), cast on all bodies from 6-Point to 12-Point, in em, 2-em and 3-em widths.

. . . .

Standard Line Hyphen Leaders No. 4, cast on all bodies from 5-Point to 12-Point, in en, em, 1½-em, 2-em and 3-em widths.

- -- --- -- -- ------ ---- --- -- -

»»›‹‹«‹

Standard Line Leaders on any body will line with every Roman or display face cast by us on that body. Any Standard Line Leader, no matter what body, may be made to line accurately with all faces on larger or smaller bodies by means of simple justification with leads and slugs; this makes them specially available for large-face date lines, etc.

The following specimen illustrates how our four styles of 10-Point Standard Line Leaders line with various faces cast on 10-Point body:

Roman No. 20.........**Ext. Old Style**........Old Style No. 9

Roman No. 23........**Woodward**......Old Style No. 10

German No. 1_____**Gothic No. 1**____French Old Style

Italic No. 20 . . . . . Gothic No. 6 . . . *O. S. Italic No. 9*

*Caledonian Italic*_____1234567890_____*French O. S. Italic*

Antique No. 1......*Gothic Italic No. 1*........Latin Series

Latin Antique. ....𝔗𝔲𝔡𝔬𝔯 𝔅𝔩𝔞𝔠𝔨........Latin Condensed Series

No. 20 Series for Newspapers

6½-POINT ROMAN No. 20

STANDARD LINING SYSTEM

One of the most important changes is our lining system. One glance at the specimen sheets issued during the past few years will show that a constant demand for something of this kind has led to ever recurrent attempts to solve the problem; but these efforts have been sporadic and inconsistent, and the failure to take into account all the conditions has rendered the results unsatisfactory. All our type is east "Standard" Line, including Romans, Italics and all job faces, therefore, all faces on one body line with one another perfectly. The advantages of this system are so many that it would be difficult to enumerate all of them. Among those that could be mentioned are: That it is now possible to line any Italic or Title with any Roman; to use heavy job letter, figures or characters with different faces on the same job, as in railroad work; to have but one set of figures in German offices where Roman is also used; that but one lot of leaders is required for each body, etc., etc. Not only are all faces of each body on the same line, but faces of different bodies justify in line with one another by the use of 2-Point or 1-Point leads, the latter being necessary only on the smaller sizes. As the spaces of all bodies are point set, fractions or multiples of points, they can be used for this justification in job work, enabling the compositor to use the caps of the next smaller size for small caps, thus resulting in a great saving of material. Not only will all faces line with the

12345 abcdefghijklmnopqrstuvwxyz 67890
ABCDEFGHIJKLMNOPQRSTUVWXYZ&
12345 ABCDEFGHIJKLMNOPQRSTUVWXYZ& 67890

One of the most important changes is our lining system. A glance at the specimen sheets issued during

---

6-POINT ROMAN No. 20

STANDARD LINING SYSTEM

One of the most important changes is our lining system. A glance at specimen sheets issued during the past few years shows that the constant demand for something of this kind has led to ever recurrent attempts to solve the problem; these efforts have been sporadic and inconsistent, however, failure to take into account all the conditions rendering the results unsatisfactory. All our type, including the Romans and Italics, Titles, Gothics, Antiques, and all job faces, are east "Standard" Line, therefore all faces on one body line with one another. The advantages of this system are so many it would be difficult to enumerate each of them. Amongst those which could be mentioned are: That it is possible now to line any Italic or Title with any Roman; to use heavy job letter, figures or characters with any body letter, as in railroad work; to have but one set of leaders in German offices where Roman is also used; but one lot of leaders will be required for each body, etc., etc. Not only are all the faces of each body on the same line, but faces of different bodies justify in line with one another by the use of 2-Point or 1-Point leads, the latter being required only on the smaller bodies. As the spaces of all our bodies are joint set, fractions or multiples of points, they can be used for justification equally as well. This feature is of the greatest importance in job work, by enabling compositor to use the caps of

12345 abcdefghijklmnopqrstuvwxyz 67890
ABCDEFGHIJKLMNOPQRSTUVWXYZ&
12345 ABCDEFGHIJKLMNOPQRSTUVWXYZ& 67890

One of the most important changes is our lining system. A glance at specimen sheets issued during

---

7-POINT ROMAN No. 20

STANDARD LINING SYSTEM

One of the most important changes is our lining system. A glance at specimen sheets issued during the past few years will show that a constant demand for something of this character has led to ever recurrent attempts to solve the problem; but these efforts have been sporadic and inconsistent, and failure to take into consideration all the conditions has rendered the results unsatisfactory. All our type, including Romans, Italics, and all job faces, are east "Standard" Line, therefore all letters on the same body line with one another perfectly. The advantages of this system are many and it would be difficult to enumerate all of them. We could mention among other things, that: It is possible by this system to line any Italic or Title with any Roman; to use heavy job letter, figures or characters in conjunction with different faces on the same job, as railroad work; but one lot of leaders is required for each body, etc., etc. Not only are faces of each body on the same line, but faces of different bodies justify in line with one another by the use of 2-Point and 1-Point

12345 abcdefghijklmnopqrstuvwxyz 67890
ABCDEFGHIJKLMNOPQRSTUVWXYZ&
12345 ABCDEFGHIJKLMNOPQRSTUVWXYZ& 67890

One of the most important changes is our new lining system. A glance at the various

---

No. 20 Series for Newspapers

8-Point Roman No. 20

STANDARD LINING SYSTEM

One of the most important changes is our lining system. A glance at specimen sheets issued during recent years clearly shows a constantly increasing demand for something of this description, which has led to ever recurrent attempts to solve the problem; these efforts have been sporadic and inconsistent, however, and failure to take into account all the conditions has rendered the results unsatisfactory. All our types are "Standarl" Line, including Romans, Italics and all job faces, therefore the faces of all letters on same body line together perfectly. It would be difficult to enumerate the many advantages of this system, but we can mention among others that: It is now possible to line any Italic or Title with any Roman; to have but one set of figures in the German offices where Roman faces are also used; but one lot of

1234 abcdefghijklmnopqrstuvwxyz 5678
ABCDEFGHIJKLMNOPQRSTUVWXYZ
12345 ABCDEFGHIJKLMNOPQR 67890

One of the most important changes is our lining system. A glance at specimen sheets

9-Point Roman No. 20

STANDARD LINING SYSTEM

One of the most important changes is our improved lining system. A glance at specimen sheets issued during recent years will show that a constant demand for something of this character has led to ever recurrent attempts to solve the problem; but these efforts have been sporadic and inconsistent, as neglect to take into account all the conditions has rendered the results unsatisfactory. All our type, including Romans, Italics and all job faces, is "Standard" Line, and all faces on same body will therefore line with one another. The advantages of this system are many and it would be difficult to enumerate all of them, but we could mention among other things that: Any Italic or Title will line with

abcdefghijklmnopqrstuvwxyz
ABCDEFGHIJKLMNOPQRSTUVW
12345 ABCDEFGHIJKLMNOPQR 67890

One of the most important changes is our new lining system. A glance at the

10-Point Roman No. 20

STANDARD LINING SYSTEM

Our lining system is certainly one of the greatest improvements in the production of type faces. A glance at the specimen sheets issued during recent years will show that an ever increasing demand for something of this description has led to repeated attempts to solve the problem; but these efforts have been sporadic and inconsistent, and failure to consider all the conditions has rendered the results unsatisfactory. Our type is cast "Standard" Line, and all faces on same body line with one another perfectly. The advantages of this system are so many that it would be

abcdefghijklmnopqrstuvwxyz
ABCDEFGHIJKLMNOPQRSTU
12345 ABCDEFGHIJKLM 67890

Our lining system is certainly one of the greatest improvements in the

ST. LOUIS, MO., U.S.A.

Unless Otherwise Ordered, En Set Figures are Furnished | THE INLAND TYPE FOUNDRY, ST. LOUIS | German, French, Spanish and Swedish Accents are Made

5-POINT ROMAN NO. 23

STANDARD LINING SYSTEM

One of the most important changes is our lining system. A glance at specimen sheets issued during the past few years shows that a constant demand for something of this kind has led to ever recurrent attempts to solve the problem; these efforts were sporadic and inconsistent and failure to take into account all the conditions has rendered the results unsatisfactory. Our type is cast "Standard" Line and includes Romans, Italics and all job faces; and therefore, faces on one body line with one another perfectly. The advantages of this system are now so many that it would be difficult to enumerate all of them. Among those that could be mentioned are to use heavy job letter, figures or characters with different faces on the same job, as in railroad work; to have but one set of figures in German offices where Roman is also used and that but one lot of leaders is required for each body. Not only are all faces of each body on the same line, but faces of another body, just Point leads; the latter being necessary only on the smaller sizes. As the spaces of bodies are point set, fractions or multiples of points, they can be used for this justification as well. This feature is of the utmost importance in job work, enabling the compositor to use the caps of next smaller size for small caps, thus resulting in a great saving of material. Not only will all faces line with the standard leaders, but the line having been placed in such a position on the body that the face will line in every case with 2-Point dotted or single rule by the use of 2-Point or 1-Point leads

12345 abcdefghijklmnopqrstuvwxyz 67890
ABCDEFGHIJKLMNOPQRSTUVWXYZ&
12345 ABCDEFGHIJKLMNOPQRSTUVWXYZ 67890

One of the important changes is our new lining system. A glance at the specimen sheets issued in recent

5½-POINT ROMAN NO. 23

STANDARD LINING SYSTEM

One of the most important changes is our lining system. In glancing over specimen sheets issued during the past few years you will notice that the constant demand for something of this kind has led to ever recurrent attempts to solve the problem; but these efforts have been sporadic and inconsistent and failure to take into account all the conditions has rendered the results unsatisfactory. All our type is cast "Standard" Line, including Romans Italics and all job faces; therefore, all faces on one body line with one another perfectly. The many advantages of this system are so great that it would be difficult to enumerate all of them. Among those that could be mentioned are: It is now possible to line any Italic or Title with all Roman; to use heavy job letter, figures or characters with different faces on the same job, as in railroad work; to have but one set of figures in German offices where Roman is also used; that but one lot of leaders is required for each body; etc., etc. Not only are all faces of each body on the same line, but faces of different bodies justify in line with one another by the use of 2-Point or 1-Point leads, the latter being necessary only on the smaller sizes. As the spaces of all bodies are point set, fractions or multiples of points, they can be used for this justification as well. This feature is of the utmost importance in job work, enabling the compositor to use the caps of the next smaller size for small caps, thus resulting in a great saving of material. Not only will all faces line with the

12345 abcdefghijklmnopqrstuvwxyz 67890
ABCDEFGHIJKLMNOPQRSTUVWXYZ&
12345 ABCDEFGHIJKLMNOPQRSTUVWXYZ 67890

One of the most important changes is our lining system. A glance at specimen sheets issued during

6-POINT ROMAN NO. 23

STANDARD LINING SYSTEM

One of the most important changes is our lining system. Glancing at specimen sheets issued during the past few years shows that a constant demand for something of this kind has led to ever recurrent attempts to solve the problem; but these efforts have been sporadic and inconsistent, failure to take into account all the conditions having rendered the results anything but satisfactory. All our type, which includes Romans, Italics, Titles, Antiques and Gothics, and all the other job faces, is cast on our "Standard" Line, and therefore, all faces of one body line with one another. The many advantages of this system make it difficult to enumerate all of them. Amongst those which may be mentioned are: That it is now possible to line any Italic or Title with any Roman; to use heavy job letter, characters or figures with different faces on the same job, as in railroad work; in German offices, where Roman is also used, to have but one set of figures; that but one lot of leaders need be purchased for each body, etc. Not only are all faces of each body on the same line, but faces of different bodies justify easily in line with one another by use of 2-Point or 1-Point leads, latter size being necessary only on smaller sizes. Spacers of all bodies are point set, fractions or multiples of

12345 abcdefghijklmnopqrstuvwxyz 67890
ABCDEFGHIJKLMNOPQRSTUVWXYZ&
12345 ABCDEFGHIJKLMNOPQRSTUVWXYZ 67890

One of the important changes is our lining system and a glance at specimen sheets issued

INLAND TYPE FOUNDRY

ST. LOUIS, MO., U. S. A.

No. 23 Series for Newspapers

8-POINT ROMAN No. 23

STANDARD LINING SYSTEM

One of the most important changes is our lining system. One glance at the specimen sheets issued during the past few years will show that the demand for something of this kind has led to ever recurrent efforts towards a solution of the problem. These efforts have been sporadic and inconsistent, and failure to take into account all the conditions has rendered the results anything but satisfactory. All our type, including job faces, is cast "Standard" Line, all faces on the one body lining together perfectly. It would be difficult to here enumerate the many advantages of this system, but we could mention among other things, that: It is now possible to line any Italic or Title letter with

123 abcdefghijklmnopqrstuvwxyz 456
ABCDEFGHIJKLMNOPQRSTUVW
12345 ABCDEFGHIJKLMNOP 67890

One of the most important changes is our lining system. A glance at the specimens

INLAND TYPE FOUNDRY

8-POINT ROMAN No. 23

STANDARD LINING SYSTEM

One of the most decided changes is our new lining system. A glance at the specimen sheets issued during the past few years will show that a constant demand for something of this kind has led to ever recurrent attempts to solve the problem; but these efforts have been sporadic and inconsistent, failure to consider and take into account all the conditions having rendered the result anything but satisfactory. All our types are cast "Standard" Line, including all Romans, Italics, Titles, Gothics and all other job faces; therefore, faces on one body will all line with one another perfectly. The advantages

abcdefghijklmnopqrstuvwxyz
ABCDEFGHIJKLMNOPQRSTU
12345 ABCDEFGHIJKLMNO 67890

One of the most important changes is our new lining system. A glance

10-POINT ROMAN No. 23

STANDARD LINING SYSTEM

One of the important changes is our lining system. A glance at the specimens issued during the past few years shows that a constant demand for something of this kind has led to recurrent attempts to master this difficult subject; the efforts were sporadic and inconsistent, and failure to take into account all conditions has rendered the result anything but satisfactory. All our type is cast "Standard" Line, including all Romans, Italics, Gothics and Titles, as well as all other faces

abcdefghijklmnopqrstuvwxyz
ABCDEFGHIJKLMNOPQRS
12345 ABCDEFGHIJKLMNO 67890

One of the greatest changes is our lining system. A glance at

ST. LOUIS, MO., U. S. A.

No. 22 Series for Newspapers

5½-Point Roman No. 22

STANDARD LINING SYSTEM

One of the most important changes is our lining system. A glance at the specimen sheets issued during the past few years shows that a constant demand for something of this kind has led to ever recurrent attempts to solve the problem. These efforts have been sporadic and inconsistent, and failure to take into account all the conditions has rendered the results unsatisfactory. Our types are cast "Standard" Line, including Romans and Italics, Titles, Gothics, Antiques, and all other job letter, consequently all faces on one body line with one another. The advantages of this system are many, and it would be difficult to enumerate all of them. We could, however, mention amongst other things, that: It is now possible to line any Italic or Title with any Roman; to use heavy job letter, figures or characters with different faces on the same job, as in railroad work; to have but one set of figures in German offices where Roman is also used, etc., etc. Not only are all the faces of each body on the same line, but faces of other bodies justify in line with one another by the use of 2-Point and 1-Point leads, it being necessary to use the latter only on the smaller bodies. As the spaces of all our bodies are point set, fractions or multiples of points, they can be used equally well for this justification. This feature is of greatest importance in job work, enabling the compositor to use the caps of the next smaller size of series for small caps, thus saving much material. Not

12345 abcdefghijklmnopqrstuvwxyz 67890
ABCDEFGHIJKLMNOPQRSTUVWXYZ&
12345 ABCDEFGHIJKLMNOPQRSTUVWXYZ& 67890

One of the most important changes is our lining
system. A glance at specimen sheets issued during

INLAND TYPE FOUNDRY

6-Point Roman No. 22

STANDARD LINING SYSTEM

One of the most important changes is our lining system. A glance at specimen sheets issued during the past few years shows that the constant demand for something of this kind has led to recurrent attempts to solve the problem. These efforts have ever been sporadic and inconsistent, and failure to take into account all the conditions rendered the results unsatisfactory. All our type, which includes Romans, Italics, Titles, and all job faces, is on "Standard" Line, and therefore all faces of one body line with one another. The advantages of this new system are so many that it would be difficult to enumerate all of them. Among them can be mentioned: That it is now possible to line any Italic or Title with any Roman; to use heavy letter, figures or characters with different faces on the same job, as railroad work; to have but one set of figures in German offices where Roman is also used; that but one lot of leaders are required for each body. Not only are all faces of each body on the same line, but faces of different bodies justify in line with one another by the use of 2-Point or 1-Point leads; the latter are necessary only on the smaller sizes. As the spaces of all bodies are point set, fractions

12345 abcdefghijklmnopqrstuvwxyz 67890
ABCDEFGHIJKLMNOPQRSTUVWXYZ&
12345 ABCDEFGHIJKLMNOPQRSTUVWXYZ& 67890

One of the most important changes is our new
lining system. A glance at the specimen sheets

7-Point Roman No. 22

STANDARD LINING SYSTEM

One of the most important changes is our new lining system. A glance at the specimen sheets issued during the past few years shows that the constant demand for something of this kind will show that the constant demand for something of this kind has led to ever recurrent attempts to solve the problem but these efforts have been sporadic and inconsistent, and the failure to take into account all the conditions has rendered the results unsatisfactory. All our type is cast on "Standard" Line, including all job faces, Romans and Italics; all faces on one body lining with one another. The advantages of this system are so many that it would be difficult to enumerate all of them. Amongst those which can be mentioned are: It is now possible to line any Italic or Title with any Roman and to use heavy job letter, characters or figures with different faces on the same job, as in railroad work; to have but one set of figures in German offices where Roman is used; that but one lot of leaders

1234 abcdefghijklmnopqrstuvwxyz 5678
ABCDEFGHIJKLMNOPQRSTUVWXYZ
12345 ABCDEFGHIJKLMNOPQRSTUVW 67890

One of the most important changes is our
lining system. A glance at the specimens

22   8-Point size shown on page 26.   St. Louis, Mo., U. S. A.

No. 24 Series for Newspapers

6-POINT ROMAN No. 24*

STANDARD LINING SYSTEM

One of the most important changes is our new system of lining. A glance through the specimen books issued during the past few years will show that a constant demand for something of this kind has led to ever recurrent attempts to solve the problem; but these efforts have been sporadic and inconsistent, and failure to take into account all the conditions has rendered the results most unsatisfactory. All our type, including Romans and Italics, Titles, Antiques, Gothics, and all job faces, is cast on "Standard" Line, and therefore all faces of one body line with one another. The advantages of this system are so abundant that it would be difficult to enumerate all of them. But among those that may be mentioned are: That it is now possible to line any Italic or Title with any Roman; to use heavy letter, figures or characters with different characters on the same job, as in railroad work; to have but one set of figures in German offices where Roman is also used; that but one set of leaders is required for each body; but one set of leaders is required for each body, the latter being needed only for the smaller bodies. Not only are all faces of various bodies justify in line with one another by the means of 2-Point and 1-Point leads, the latter being needed only for the smaller bodies. As the spaces of all our bodies are point set, fractions or multiples of

12345 abcdefghijklmnopqrstuvwxyz 67890
ABCDEFGHIJKLMNOPQRSTUVWXYZ&
ABCDEFGHIJKLMNOPQRSTUVWXYZ&

One of the important changes is our new system of lining. A glance at the specimen sheets that issued

7-POINT ROMAN No. 24*

STANDARD LINING SYSTEM

One of the most important changes is our new lining system. A glance at the specimen sheets issued during the past few years will show that the constant demand for something of this kind has led to ever recurrent efforts to solve the problem; but these attemps have been sporadic and inconsistent, the failure to take into account all the conditions having rendered the results unsatisfactory. All our type, including Romans and Italics, as well as Titles, Antiques, Gothics, and all other job faces, is cast "Standard" Line; therefore all faces of one body line with one another. The advantages of this system are so many that it would be very difficult to enumerate all of them. Among those that may be mentioned are: It is now possible to line any Italic or Title with any Roman; to make use of heavy job letter, figures or characters with different faces on the same job, as in railroad work; to have but one set of figures in German offices where Roman is also used; that but one lot of leaders is needed for each body, etc. Not only are all faces of each body on the same

12345 abcdefghijklmnopqrstuvwxyz 67890
ABCDEFGHIJKLMNOPQRSTUVWXYZ&
ABCDEFGHIJKLMNOPQRSTUVWXYZ&

One of the important changes is our system of lining. A glance at the specimen sheets that

8-POINT ROMAN No. 24

STANDARD LINING SYSTEM

One of the most important changes is our new system of lining. One glance at the specimen sheets issued during the past few years shows that the constant demand for something of this nature has led to ever recurrent attempts to solve the problem; but these efforts have been sporadic and inconsistent, and failure to take into account all the conditions has rendered the results unsatisfactory. All our type, including Romans, Italics, and Titles, Antiques, Gothics, as well as all other faces, is cast "Standard" Line, by reason of which all faces of one body will line with one another. The advantages of this system are so many that it would be difficult to enumerate all of them; but among those that may here be mentioned are: It is now possible to line any Italic or Title with any Roman; to use heavy job letter, figures or characters of various 12345 abcdefghijklmnopqrstuvwxyz 67890 ABCDEFGHIJKLMNOPQRSTUVWXYZ&

ABCDEFGHIJKLMNOPQRSTUVWXYZ&
One of the important changes is our new system of lining. A glance at the specimen

No. 25 Series for Newspapers

5-POINT ROMAN No. 25

STANDARD LINING SYSTEM

One of the most important changes is our lining system. A glance at the specimen sheets issued during the past few years shows that a constant demand for something of this kind has led to ever recurrent attempts to solve the problem; these efforts, however, have been sporadic and without consistency, and failure to take into account all the conditions has rendered the results highly unsatisfactory. All our type, including Romans and Italics, Titles, Antiques, Gothics, and all job faces, is cast on "Standard" Line; therefore, all faces cast on the same body line with one another perfectly. The advantages of this system are so many that it would be difficult to enumerate all of them. But among those that could be mentioned are: That it is now possible to line any Italic or Title with any Roman; to use heavy job letter and figures or characters with different faces on the same job, as in railroad work; to have but one set of figures in German offices where Roman is also used; that but one set of leaders is required for each body, etc. Not only are all the faces of each body, but the different faces of the same line, with one another by the use of 2-Point and 1-Point leads. As the spaces of all bodies are point set, or fractions or multiples of points, they can also be used for this justification. This feature is of the very highest importance in job work, enabling the compositor to use the caps of the next smaller size for small caps, resulting in a great saving of material. Not only will all faces line with the standard leaders

12345 abcdefghijklmnopqrstuvwxyz 67890

ABCDEFGHIJKLMNOPQRSTUVWXYZ&

12345 ABCDEFGHIJKLMNOPQRSTUVWXYZ& 67890

One of the important changes is our lining system. A glance at the specimen sheets issued during the past few

5½-POINT ROMAN No. 25

STANDARD LINING SYSTEM

One of the most important changes is our lining system. A glance at specimen sheets issued during the past few years will show that a constant demand for something of this character has led to ever recurrent attempts to solve the problem; but these efforts have been sporadic and inconsistent, and failure to take into consideration all the conditions has rendered the results unsatisfactory. All our type, including Romans, Italics, and all job faces, is cast "Standard" Line; therefore all faces on the same body line with one another perfectly. The advantages of this system are so many that it would be difficult to enumerate all of them. We could mention among other features, that: It is possible by this system to line any Italic or Title with any Roman; to use heavy job letter, figures or characters in conjunction with different faces on the same job, as in railroad work; but one lot of leaders is required for each body, etc. Not only are the faces of each body on the same line, but faces of different bodies justify in line with leads, the latter being necessary only on the use of 2-Point and 1-Point smaller sizes. As the spaces of all bodies are point set, fractions or multiples of points, it is possible to use them for this justification as well. This feature is of the utmost value in job work, enabling the compositor to use

12345 abcdefghijklmnopqrstuvwxyz 67890

ABCDEFGHIJKLMNOPQRSTUVWXYZ&

12345 ABCDEFGHIJKLMNOPQRSTUVWXYZ 67890

One of the important changes is our lining system. A glance at the specimen sheets issued

6-POINT ROMAN No. 25

STANDARD LINING SYSTEM

One of the most important changes is our new lining system. A glance at the specimen sheets issued during the past few years will show one that the constant demand for something of this nature has led to ever recurrent attempts to solve the problem; but these efforts have been sporadic and inconsistent, and failure to take into account all the conditions has rendered the results unsatisfactory. All our type, including Romans, Italics, and all job faces, is cast "Standard" Line, and therefore all faces of one body line with one another. The advantages of this new system are so many that it is difficult to enumerate them all. Among those which can be mentioned are: it is now possible to line accurately any Italic or Title with any Roman; to use heavy letter, figures or characters with different faces on the same job, as in railroad work; to have but one set of figures in German offices where Roman is used; that but one lot of leaders are required for each body. Not only are all faces of each body on the same line, but the faces of different bodies justify in line with one another by the use of 2-Point and 1-Point leads, the latter being necessary

12345 abcdefghijklmnopqrstuvwxyz 67890

ABCDEFGHIJKLMNOPQRSTUVWXYZ&

12345 ABCDEFGHIJKLMNOPQRSTUVWXYZ 67890

One of the important changes is our new lining system. A glance at the specimen

ST. LOUIS, MO., U. S. A.

No. 27 Series—On Point and Half-Point Sets

6-POINT ROMAN No. 27

STANDARD LINING SYSTEM

One of the most important changes is our lining system. One glance at the specimen sheets issued during the past few years will show that a constant demand for something of this kind has led to ever recurrent efforts to solve the problem; but these attempts have been sporadic, also inconsistent, and failure to take into account every condition has rendered the results unsatisfactory. All our type, including Romans and Italics, and Titles, Antiques, Gothics, and all other job faces, is cast on "Standard" Line; therefore all faces on one body line perfectly with one another. The advantages of this system are so many that it would be a difficult matter to enumerate all of them. But among those which may be mentioned are: That is is now possible to line any Italic or Title with any job letter, figures or characters with different faces on the same job, as in railroad work; to have but one set of figures in German offices where Roman is also used; that but one lot of leaders need be purchased for each body, etc. Not only are all the faces of each body, on the same line, but the faces of different bodies justify line with one another readily by the use

12345 abcdefghijklmnopqrstuvwxyz 67890
ABCDEFGHIJKLMNOPQRSTUVWXYZ&
12345 ABCDEFGHIJKLMNOPQRSTUVWXYZ 67890

One of the most important changes is our new lining system. A glance at the specimen sheets

INLAND TYPE FOUNDRY

7-POINT ROMAN No. 27

STANDARD LINING SYSTEM

One of the most important changes is our new lining system. A glance at the specimen sheets issued during the past few years will show that a constant demand for something of this nature has led to ever recurrent efforts to solve the vexing problem. But these attempts have been sporadic and inconsistent, the failure to take into account all the conditions has rendered the results unsatisfactory. All our type, including Romans, Italics, and Titles, Antiques, Gothics, and all other faces, is cast on "Standard" Line, and therefore all faces of one body will line with one another. The advantages of this system are so many that it would be difficult to enumerate them all. But among those which may be mentioned are: That it is now possible to line any Italic or Title with any Roman; to use heavy job letter, figures or characters with different faces on the same job, as in railroad work; to have but one set of

123 abcdefghijklmnopqrstuvwxyz 456
ABCDEFGHIJKLMNOPQRSTUVW
12345 ABCDEFGHIJKLMNOPQRSTUV 67890

One of the most important changes is our new lining system. A glance at specimen

8-POINT ROMAN No. 27

STANDARD LINING SYSTEM

One of the most important changes is our lining system. A glance at the specimen sheets issued during recent years shows that a constant demand for something of this nature has led to ever recurrent efforts to solve the vexing problem; but these attempts were sporadic and inconsistent, and the failure to take into account all the conditions has rendered results unsatisfactory. All our type, which includes Romans, Italics, Titles, and Antiques, Gothics, and all other job faces, is cast "Standard" Line, hence all faces on each body line with one another perfectly. The advantages of this system are so numerous that it would be impossible to count them all. Among those that may here be mentioned are: It is now possible to

12 abcdefghijklmnopqrstuvwxyz 34
ABCDEFGHIJKLMNOPQRSTUV
12345 ABCDEFGHIJKLMNOPQRS 67890

One of the most important changes is our new lining system. A glance over

ST. LOUIS, MO., U. S. A.

Various Faces for Newspapers

8-POINT ROMAN No. 22

STANDARD LINING SYSTEM

One of the most important changes is our lining system. A glance at the specimen sheets issued during the last few years shows that constant demand for something of this kind has led to ever recurrent attempts to solve the problem; but these efforts have been sporadic and inconsistent, and failure to take into account all the conditions has rendered the results anything but satisfactory. All our type, including Romans, Italics, Titles, Antiques and Gothics, and all other jobbing faces, is cast "Standard" Line, and therefore all faces of one body will line with one another. Advantages of this system are so many that it would be difficult to enumerate all of them. Amongst those which may be stated are: That

12 abcdefghijklmnopqrstuvwxyz 34
ABCDEFGHIJKLMNOPQRSTUVW&
12345 ABCDEFGHIJKLMNOP 67890

*One of the most important changes is
our new lining system. A glance at the*

6-POINT ROMAN No. 25

STANDARD LINING SYSTEM

One of the most important changes is our lining system. A glance at the specimen sheets issued during the past few years shows that a constant demand for something of this kind has led to recurrent attempts to solve the problem. These efforts have been sporadic and inconsistent, the failure to take into account all the conditions having rendered the results obtained anything but satisfactory. All our type, which includes Romans, Italics, Titles, Antiques, and all jobbing faces, is cast on "Standard" Line; therefore, all faces of one body line with one another. The advantages of this system are so many that it would be difficult to enumerate all of them. Among those that could be mentioned are: It is possible now to line any Italic or Title with any Roman; to use heavy job letter, figures or characters with different faces on the same job, as in railroad work; to have but one set of figures in German offices where Roman type is also used; that but one lot of leaders is required for each body in an office, etc. Not only are all faces of each body on the same line, but faces of different bodies line with one another with easy justification by use of 2-Point or 1-Point leads, the latter being required only on the smaller sizes. As the spaces of all

12345 abcdefghijklmnopqrstuvwxyz 67890
ABCDEFGHIJKLMNOPQRSTUVWXYZ&
12345 ABCDEFGHIJKLMNOPQRSTUVWXYZ& 67890

*One of the most important changes is our lining
system. A glance at specimen sheets issued during*

9-POINT ROMAN No. 24

STANDARD LINING SYSTEM

One of the most important changes is our lining system. A glance at the specimen sheets issued during years past shows that the constant demand for something of this kind has led to ever recurrent attempts to solve the problem; but these efforts have been sporadic and inconsistent, and failure to take into account all the conditions has rendered the results anything but satisfactory. All our type, including Romans, Italics, Titles, Gothics, and all other jobbing faces, is cast on the improved "Standard" Line, therefore all faces on one body line with one another. It would be very difficult to here enumerate the many advantages

abcdefghijklmnopqrstuvwxyz
ABCDEFGHIJKLMNOPQRSTUVW
12345 ABCDEFGHIJKLM 67890

*One of the important changes is our
lining system. A glance into specimen*

ST. LOUIS, MO., U.S.A.

BOOK  FACES

5-POINT ROMAN NO. 20

THE STANDARD LINING SYSTEM

One of the most important changes is our lining system. A glance at the specimen sheets issued during past few years shows that constant demand for something of this kind has led to ever recurrent attempts to solve the problem; but these efforts have been sporadic and inconsistent, and failure to take into account all the conditions has rendered the results anything but satisfactory. All our type is cast "Standard" Line, including Romans and Italics, Titles, Antiques, Gothics, and all other job faces; therefore all faces of one body line with one another. The advantages of this system are so many that it would be difficult to enumerate all of them. Amongst those which can be mentioned are: That it is now possible to line any Italic or Title with any Roman; to use heavy job letter, figures or characters with different faces on the same job, as in railroad work, as in railroad work; to have but one set of figures in German offices where Roman is also used; that but one lot of leaders need be purchased for each body, etc., etc. Not only are all the faces of each body on the same line, but faces of different bodies justify in line with one another perfectly by the use of 2-Point or 1-Point leads, the latter being necessary only on the smaller bodies. As spaces of all bodies are point set, fractions or multiples of points, they can be used for this justification as well. This feature is of the greatest importance in all job work, and by enabling the compositor to use the caps of the next smaller size of series for small caps results in the saving of much material. Not only will all faces line with the standard leaders, but will line with 2-Point single or dotted rule perfectly by the use of 2-Point or 1-Point leads and standard quads. In spite of the general use of leaders it is often necessary to use rule, and this innovation, which enables the printer to have accurate lining without the use of cardboard, etc., will be appreciated. It is not only doing away with the useless expense of the I combinations, but securing absolute immunity from breakage after the type is locked up, and insuring absolute electrotypes. The same plan is followed with the descenders, making all non-kerning. Throughout the Old Styles the figures above and below the line have been abandoned, and while retaining the old style design these important characters have been made of uniform size and line, adapting them to all uses and making them handsomer and more legible. All small

12345 abcdefghijklmnopqrstuvwxyz 67890
ABCDEFGHIJKLMNOPQRSTUVWXYZ&
12345 ABCDEFGHIJKLMNOPQRSTUVWXYZ& 07890

One of the most important changes is our new lining system. A glance at the specimen sheets issued during the past few years shows that a constant demand

INLAND TYPE FOUNDRY

5½-POINT ROMAN NO. 20

THE STANDARD LINING SYSTEM

One of the most important changes is our lining system. Now a glance at the specimen sheets issued during the past few years shows that the constant demand for something of this kind has led to ever recurrent attempts to solve the problem; but these efforts have been sporadic and inconsistent, failure to take into account all the conditions rendering results unsatisfactory. All our type is cast on the improved "Standard" Line, including Romans and Italics, Titles, Antiques, Gothics, and all other faces; therefore all faces of one body line with one another. The advantages of this system are so many that it would be difficult to enumerate all of them. Amongst those which may be mentioned are: That it is now possible to line any Italic or Title with any Roman; to use heavy job letter, figures or characters with different faces on the same job, as in railroad work; to have but one set of figures in German offices where Roman is also used; that but one lot of leaders need be purchased for each body, etc. etc. Not only are all the faces of each body on the same line, but faces of different bodies justify in line with one another accurately by the use of 2-Point or 1-Point leads, the latter being necessary only on the smaller bodies. As the spaces of all the bodies are point set, fractions or multiples of points, they can be used for this justification as well. This feature is of the greatest importance in all job work, and by enabling the compositor to use the caps of the next smaller size of the series for small caps results in the saving of much material. Not only will all faces line with the standard leaders, but will line with 2-Point single or dotted rule perfectly by the use of 2-Point or 1-Point leads and standard quads. In spite of the general use of leaders it is often necessary to use rule, and this innovation, which enables the printer to have accurate lining without the use of cardboards, will be appreciated. Throughout the Old Styles the figures above and

12345 abcdefghijklmnopqrstuvwxyz 67890
ABCDEFGHIJKLMNOPQRSTUVWXYZ&
12345 ABCDEFGHIJKLMNOPQRSTUVWXYZ& 67890

One of the most important changes is our lining system. A glance at the specimen sheets issued during the past few years will show that the

ST. LOUIS, MO., U. S. A.

BOOK FACES

6-POINT ROMAN NO. 20

THE STANDARD LINING SYSTEM

One of the most important changes is our lining system. A glance at the specimen sheets issued during the past few years will show that the constant demand for something of this kind has led to ever recurrent attempts to solve the problem; but these efforts have been sporadic and inconsistent, failure to take into account all the conditions rendering results unsatisfactory. All our type is cast on the improved "Standard" Line, including Romans and Italics, Titles, Antiques, Gothics, and all other job faces; therefore all faces of one body line with one another. The advantages of this system are so many that it would be difficult to enumerate all of them. Amongst those which can be mentioned are: That it is now possible to line any Italic or Title with any Roman; to use heavy job letter, figures or characters with different faces on the same job, as in railroad work; to have but one set of figures in German offices where Roman is also used; that but one all the faces of each body on the same line, but faces of different bodies justify in line with one another by the use of 2-Point or 1-Point leads, the latter being necessary only on the smaller bodies. As the spaces of all bodies are point set, fractions or multiples of points, they can be used for this justification as well. This feature is of the greatest importance in job work, and by enabling the compositor to use the caps of the next smaller size of the series for small caps results in the saving of much material. Not only will all faces line with the standard leaders, but will line with 2-Point single or dotted rule perfectly by the use of 2-Point or 1-Point leads and standard quads. In spite of the general use of leaders it is often necessary to use rule, and this innovation, which enables the

12345 abcdefghijklmnopqrstuvwxyz 67890
ABCDEFGHIJKLMNOPQRSTUVWXYZ&
12345 ABCDEFGHIJKLMNOPQRSTUVWXYZ& 67890

One of the important changes is our lining system. A glance at the specimen sheets issued during the past few years shows that the

7-POINT ROMAN NO. 20

THE STANDARD LINING SYSTEM

One of the important changes is our lining system. A glance at the specimen sheets issued during the past few years will show that the constant demand for something of this kind has led to ever recurrent attempts to solve the problem. These efforts, however, have been sporadic and inconsistent, failure to take into account all the conditions having rendered the results unsatisfactory. All our type is cast "Standard" Line, including Romans, Italics, and all job faces, therefore all faces on the same body line with one another. The advantages of this system are so many that it would be difficult to enumerate all of them. Among those which may be mentioned are: It is now possible to line any Italic or Title with any Roman; in German offices, where Roman is also used, to have but one set of figures; to use heavy job letter, figures or characters with different faces on the same job, as in railroad work; that but one lot of leaders need be purchased for each body, etc. Not only are all the faces of each body on the same line, but faces of different bodies justify in line with one another by the use of 2-Point or 1-Point leads, the latter size being necessary only on the smaller bodies. The spaces of all bodies being point set, fractions or multiples of points, they can be used for this justification as well. This feature is of the greatest importance in job work, and by enabling the compositor to

12345 abcdefghijklmnopqrstuvwxyz 67890
ABCDEFGHIJKLMNOPQRSTUVWXYZ&
12345 ABCDEFGHIJKLMNOPQRSTUVWXYZ& 67890

One of the most important changes is our lining system and a glance at the specimen sheets issued during past few

BOOK  FACES

8-POINT ROMAN No. 20

THE STANDARD LINING SYSTEM

One of the most important changes is our new lining system. A glance at the specimen sheets will show that the constant demand for something of this kind has led to ever recurrent attempts to solve the problem. These efforts have been sporadic and inconsistent, failure to take into account all the conditions having rendered the results anything but satisfactory. All our type, including Romans and Italics, Titles, Antiques, Gothics, and all job faces, is cast "Standard" Line, and therefore all faces of one body line with one another. The advantages of this system are so many that it were difficult to enumerate all of them. Amongst those which may be mentioned are: That it is now possible to line any Italic or Title with any Roman; to use heavy job letter, characters or figures with different faces on the same job, as in railroad work; in German offices, where Roman is also used, to have but one set of figures; that but one lot of leaders need be purchased for each body, etc. Not only are all faces of each body on the same line, but

12345 abcdefghijklmnopqrstuvwxyz 67890
ABCDEFGHIJKLMNOPQRSTUVWXYZ&
12345 ABCDEFGHIJKLMNOPQR 67890

One of the most important changes is our new lining system. A glance at the specimen sheets issued during

INLAND TYPE FOUNDRY

9-POINT ROMAN No. 20

THE STANDARD LINING SYSTEM

One of the most important changes is our lining system. A glance at specimen sheets issued during the past few years shows that the constant demand for something of this kind has led to ever recurrent attempts to solve the problem; but these efforts have been sporadic and inconsistent, and failure to take into account all the conditions has rendered the results unsatisfactory. All our type, including Romans, Italics, Titles, Antiques, Gothics, and all other job faces, is "Standard" Line, and therefore all faces of one body line with one another. The advantages of this system are so many that it were difficult to enumerate all of them. Amongst those which may be mentioned are: That it is possible to line any Italic or Title with any Roman; to use heavy job letter or characters with different faces on the same job, as in railroad work; to have but one set of figures in German offices where Roman is

abcdefghijklmnopqrstuvwxyz.
ABCDEFGHIJKLMNOPQRSTUVWXYZ&
12345 ABCDEFGHIJKLMNOPQR 67890

One of the most important changes is our lining system and a glance at specimen sheets issued during

St. Louis, Mo., U. S. A.

BOOK FACES

German, French, Spanish and Swedish
Accents are made for these faces

11-POINT ROMAN No. 20

STANDARD LINING SYSTEM

One of the most important changes is our lining system. A glance at specimen sheets issued during the past few years will show that the constant demand for something of this description has led to ever recurrent efforts towards a solution of the problem; but these attempts have been sporadic and inconsistent, failure to take into account all conditions rendering the results unsatisfactory. All our types are "Standard" Line, including Romans and Italics, Titles, Gothics and all other job faces, and therefore all faces of one body line with one another. Among the

abcdefghijklmnopqrstuvwxyz
ABCDEFGHIJKLMNOPQRSTUVWXY
12345 ABCDEFGHIJKLMNOPQR 67890

One of the most important changes is our lining system. A glance at the specimen

St. Louis, Mo., U. S. A.

10-POINT ROMAN No. 20

STANDARD LINING SYSTEM

One of the most important changes is our lining system. A glance at specimen sheets issued during the past few years shows that the constant demand for something of this character has led to ever recurrent attempts to solve the problem; these efforts have been sporadic and inconsistent, however; failure to take into account all the conditions rendering results unsatisfactory. All our types are cast on the improved "Standard" Line, including all Romans, Italics, Titles, Gothics, and other job faces, therefore, all faces of one body line with one another. Among many of the other advantages of this system which may here be mentioned are: That it is now possible to line

abcdefghijklmnopqrstuvwxyz
ABCDEFGHIJKLMNOPQRSTUVWXYZ&
12345 ABCDEFGHIJKLMNOPQR 67890

One of the important changes is that of the improved lining system. A glance at specimen

INLAND TYPE FOUNDRY

BOOK FACES

12-POINT ROMAN No. 20

OUR NEW LINING SYSTEM

One of the most important changes is our lining system. A glance at the specimen sheets issued during recent years shows that a constant demand for something of this kind has led to ever recurrent attempts to solve the problem; but these efforts have been sporadic and inconsistent, failure to take into account all of the conditions having rendered the results anything but satisfactory. All our type is cast on "Standard" Line, including all job

abcdefghijklmnopqrstuvwxyz
ABCDEFGHIJKLMNOPQRSTUV
12345 ABCDEFGHIJKLMNO 67890

*One of the greatest changes is that of
our lining system. A glance at recent*

6-POINT ROMAN No. 21

THE STANDARD LINING SYSTEM

One of the most important changes is our new system of lining. A glance at the specimen sheets issued during the past few years shows that the constant call for something of this kind has led to ever recurrent attempts to solve the problem; however, these efforts have been sporadic and inconsistent, and the failure to take into account all the conditions has caused the results to be unsatisfactory. All our type, including Romans and Italics, Titles, Antiques and Gothics, as well as all the other job faces, is cast on the improved "Standard" Line, and therefore all faces of one body line with one another. The advantages of this new system are so many that it would be difficult to enumerate all of them. But among those which may be mentioned are: That it is now possible to line any Italic or Title with any Roman; to use heavy job letter, figures or characters with different faces on the same job, as in railroad work; to have but one set of figures in German offices where Roman is also used; that but one lot of leaders need be purchased for each body, etc. Not only are all the faces of each body on the same line, but faces of different bodies justify in line with one another accurately by the use of 2-Point or 1-Point leads, the latter size being necessary only on the smaller bodies. As the spaces of all our bodies are cast on point sets, fractions or multiples of points, they can be used for this justification as well. This feature is of the greatest importance in job work, and by enabling the compositor to use the caps of the next smaller size results in the saving of much material. Not only will all faces line with leaders

abcdefghijklmnopqrstuvwxyz
ABCDEFGHIJKLMNOPQRSTUVWXYZ&
12345 ABCDEFGHIJKLMNOPQRSTUVWXYZ& 67890

*One of the most important changes is our lining system. A
glance at the specimen sheets issued during the past few years*

ST. LOUIS, MO., U. S. A.

BOOK FACES

5½-POINT ROMAN No. 22

THE STANDARD LINING SYSTEM

One of the most important changes is our system of lining. A glance at the specimen books issued during the past few years shows that a constant demand for something of this kind has led to ever recurrent attempts to solve the problem; but the various efforts have been sporadic and inconsistent, and failure to take into account all the conditions has rendered the results highly unsatisfactory. All our type, including Romans and Italics, as well as Titles, Antiques, Gothics, and all our other job faces, is cast "Standard" Line; therefore all faces of one body line with one another. The advantages of this system are so many that it would be difficult to enumerate all of them. Among those which may be mentioned are: That it is now possible to line any Italic or Title with any Roman; in German offices where Roman is also used to have but one set of figures; to use heavy letter, figures or characters with different faces on the same job, as in railroad work; that but one set of leaders need be purchased for each body, etc. Not only are all the faces of each body on the same line, but faces of different bodies justify in line with one another accurately by the use of 2-Point and 1-Point leads, the latter size being necessary only with the smaller bodies. As the spaces of all bodies are point set, fractions or multiples of points, they can be used for this justification as well. This feature is of greatest importance in job work, and by enabling the compositor to use the caps of the next smaller size of the series results in the saving of much material. Not only will all faces line with the standard line with the standard leaders, but they will line accurately with 2-Point single or dotted rule by the use of 2-Point or 1-Point leads and standard quads for justification. In spite of the general use of leaders, it is very often necessary to use brass rule, and this innovation, which enables the printer to have perfect lining, will be highly appreciated. Throughout the Old Styles all the figures

12345 abcdefghijklmnopqrstuvwxyz 67890
ABCDEFGHIJKLMNOPQRSTUVWXYZ&
12345 ABCDEFGHIJKLMNOPQRSTUVWXYZ& 67890

One of the most important changes is our new system of lining. A glance at the specimen sheets issued during the past few years shows

INLAND TYPE FOUNDRY

6-POINT ROMAN No. 22

THE STANDARD LINING SYSTEM

One of the most important changes is our new system of lining. A glance at the specimen sheets issued during the past few years shows that the constant demand for something of this nature has led to ever recurrent attempts to solve the problem. These efforts, however, have been sporadic and inconsistent, the failure to take into account all the conditions having rendered unsatisfactory results. All our type is cast "Standard" Line, including Romans and Italics, Titles, Antiques, Gothics, Latins, and all other job faces; therefore all faces of one body line accurately with one another. The advantages of this system are so many that it would be difficult to enumerate all of them. Among those which can be mentioned are: That it is now possible to line any Italic or Title with any Roman; to use heavy job letter, figures or characters with different faces on the same job, as in railroad work; in German offices where Roman is also used to have but one set of figures; that but one lot of leaders need be purchased for each body in an office, etc. Not only are all the faces of each body on the same line, but faces of different bodies justify in line with one another by the use of 2-Point and 1-Point leads, the latter being necessary only on the smaller bodies. As the spaces of all bodies are point set, fractions or multiples of points, they can be used for this justification as well. This feature is of the greatest importance in job work, and by enabling compositors to use the caps of the next smaller size of the series results in the saving of a large amount of material. Not only will all faces line with the standard

12345 abcdefghijklmnopqrstuvwxyz 67890
ABCDEFGHIJKLMNOPQRSTUVWXYZ&
12345 ABCDEFGHIJKLMNOPQRSTUVWXYZ& 67890

One of the most important changes is our lining system. A glance at the specimen sheets issued during the past few years

ST. LOUIS, MO., U.S.A.

BOOK

FACES

THE STANDARD LINING SYSTEM

One of the most important changes is our system of lining. A glance at the specimen sheets issued during the past few years shows that a constant demand for something of this nature has led to ever recurrent attempts to solve the problem; but these efforts have been sporadic and inconsistent, and failure to take into account all the conditions has caused the results to be unsatisfactory. All our type, including Romans and Italics, Titles, Antiques, Gothics, and all other job faces, is cast "Standard" Line; therefore all the faces of the same body line with one another. The advantages of this system are so many that it would be impossible to enumerate all of them. Among the ones which can be mentioned are: That it is possible now to line any Italic or Title with any Roman; to use heavy letter, figures or characters with different faces on the same job, as in railroad work; to have but one set of figures in German offices where Roman is also used; that but one lot of leaders need now be purchased for each body, etc. Not only are all the faces of each body on the same line, but faces of the different bodies justify in line with one another by use of 2-Point and 1-Point leads, the latter size being necessary only on the smaller bodies. As the spaces

12345   abcdefghijklmnopqrstuvwxyz (67890)

ABCDEFGHIJKLMNOPQRSTUVWXYZ&

12345   ABCDEFGHIJKLMNOPQRSTUVWXYZ& 67890

One of the most important changes is our new lining system. A glance at the specimen sheets issued during

THE STANDARD LINING SYSTEM

One of the most important changes is our new lining system. A glance at the specimen sheets issued during the past few years shows that the constant demand for something of this kind has led to ever recurrent attempts to solve this great problem; but these efforts have been sporadic and inconsistent, the failure to take into account all the conditions having rendered results highly unsatisfactory. All our type, including Romans and Italics, Titles, Antiques, Gothics, and all the other job faces, is cast on "Standard" Line, and therefore all the faces of one body line with one another. The advantages of this system are so many that it would be difficult to enumerate all of them. Among those which may be mentioned are: That it is now possible to line any Italic or Title with any Roman; to use heavy job letter or figures and characters with different faces on the same job, as in railroad work; to have but one set of figures in German offices where Roman is also used; that but one lot of leaders is required

12345   abcdefghijklmnopqrstuvwxyz   67890

ABCDEFGHIJKLMNOPQRSTUVWXYZ&

12345   ABCDEFGHIJKLMNOPQRSTUVWXYZ& 67890

One of the most important changes is our lining system. A glance at the specimen sheets issued in

Unless otherwise ordered, the En Set
Figures are supplied with all fonts

German, French, Spanish and Swedish
Accents are made for these faces

BOOK

FACES

5-POINT ROMAN NO. 23

THE STANDARD LINING SYSTEM

One of the most important changes is our lining system. Now a glance at specimen sheets issued during the past few years shows that a constant demand for something of this kind has led to ever recurrent attempts to solve the problem; these efforts have been sporadic and inconsistent, and failure to take into account all the conditions has rendered the results unsatisfactory. All our type is cast on the improved "Standard" Line, including Romans, Italics, Titles, Antiques, Gothics, and all other job faces; therefore all faces of one body line with one another. The advantages of this system are so many that it would be difficult to enumerate all of them; but amongst those which may be mentioned are: That it is possible now to line any Italic or Title with any Roman; to use heavy job letter, figures or characters with different faces on same job, as in railroad work; but one set of figures are required in German offices where Roman is also used; but one lot of leaders need be purchased for each body, etc., etc. Not only are all the faces of each body on the same line, but faces of different bodies justify in line with one another accurately by the use of 2-Point or 1-Point leads, the latter bodies are readily set, fractions or multiples, as the spaces of all used for this justification as well. This feature is of the greatest importance in job work, and by enabling the compositor to use the caps of the next smaller size of the series for small caps results in the saving of much material. Not only will all faces line with the standard leaders, but will line with 2-Point single or dotted rule perfectly by the use of 2-Point leads and our standard quads. In spite of the general use of leaders it is often necessary to use rule, and this innovation, which enables the printer to have accurate lining without the use of cardboards, will be very highly appreciated. In the Old Styles the figures above and below the line have been abandoned, and while retaining the old style design these important characters have been made of uniform size and line, and are now adaptable for all uses, being much handsomer and more legible. All small cap sorts liable to be confused with the same

12345 abcdefghijklmnopqrstuvwxyz 67890

ABCDEFGHIJKLMNOPQRSTUVWXYZ&

12345 ABCDEFGHIJKLMNOPQRSTUVWXYZ 67890

One of the most important changes is our new lining system. A glance at the specimen sheets issued during the past few years shows that constant

INLAND TYPE FOUNDRY

5½-POINT ROMAN NO. 23

THE STANDARD LINING SYSTEM

One of the most important changes is our lining system. Now a glance at specimen sheets issued during the past few years shows that the constant demand for something of this kind has led to ever recurrent attempts to solve the problem; these efforts have been sporadic and inconsistent, and failure to take into account all the conditions has rendered the results unsatisfactory. All our type is cast on the improved "Standard" Line, including Romans and their Italics, Titles, Antiques, Gothics, and all other job faces; therefore all faces of one body line with one another. The advantages of this system are so many that it would be difficult to enumerate all of them. Amongst those which can be mentioned are: That it is now possible to line any Italic or Title with any Roman; to use heavy job letter, figures or characters with different faces on the same job, as in railroad work: to have but one set of figures in German offices where Roman is also used; that but one lot of leaders need be purchased for each body, etc., etc. Not only are all the faces of each body on the same line, but faces of different bodies justify in line with one another accurately by the use of 2-Point or 1-Point leads, the latter being necessary only on the smaller bodies. As the spaces of all bodies are point set, fractions or multiples of points they can be used for this justification as well. This feature is of the greatest importance in job work, and by enabling the compositor to use the caps of the next smaller size of the series for small caps results in the saving of material. Not only will all faces line with the standard leaders, but they will line with 2-Point single or dotted rule perfectly by the use of 2-Point or 1-Point leads and standard quads. In spite of the use of leaders it is often necessary to use rule and this innovation, which enables printers to have accurate lining without the use of cardboards, will be appreciated. Throughout the Old Styles the figures above and below the line have all been

12345 abcdefghijklmnopqrstuvwxyz 67890

ABCDEFGHIJKLMNOPQRSTUVWXYZ&

12345 ABCDEFGHIJKLMNOPQRSTUVWXYZ 67890

One of the most important changes is our new lining system. A glance at the specimen sheets issued during the past few years shows

ST. LOUIS, MO., U. S. A.

BOOK FACES

6-POINT ROMAN No. 23

THE STANDARD LINING SYSTEM

One of the most important changes is our improved lining system. A glance at the specimen sheets issued during the past few years shows that a constant demand for something of this kind has led to ever recurrent attempts to solve the problem. These efforts, however, have been sporadic and inconsistent, failure to take into account all the conditions having rendered the results unsatisfactory. All our type is cast "Standard" Line, including Romans, Italics, and all job faces, consequently all faces on one body line with one another. The advantages of this system are so many that it would be difficult to enumerate all of them. Amongst those which could be mentioned are: That it is now possible to line any Italic or Title with any Roman; to use heavy job letter or figures with different faces on the same job, as in railroad work; to have but one set of figures in German offices where Roman is also used; that but one lot of leaders is required for each body, etc. Not only are all the faces of each body on the same line, but faces of different bodies justify in line with one another by the use of 2-Point and 1-Point leads, the latter being necessary only on the smaller bodies. As the spaces of all bodies are point set, fractions or multiples of points, they can be used for this justification as well. This feature is of the greatest importance in job work, enabling the compositor to use caps of the next smaller size of the series for small caps, and thus saving much material. Not only will all faces line with the standard leaders, but the line has been placed in such position on the body that in every

12345 abcdefghijklmnopqrstuvwxyz 67890
ABCDEFGHIJKLMNOPQRSTUVWXYZ&
12345 ABCDEFGHIJKLMNOPQRST 67890

One of the most important changes is our lining system. A glance at the specimen sheets issued during the past few years

8-POINT ROMAN No. 23

THE STANDARD LINING SYSTEM

One of the most important changes is our lining system. A glance at specimen sheets issued during the past few years shows that the constant demand for something of this kind has led to ever recurrent attempts to solve the problem; but these efforts have been sporadic and inconsistent, and failure to take into account all the conditions has rendered the results unsatisfactory. All our type, including Romans, Italics, Titles, Antiques, Gothics, and all other job faces, is "Standard" Line, and therefore all faces of one body line with one another. The advantages of this system are so many that it were difficult to enumerate all of them. Amongst those which may be mentioned are: That it is possible to line any Italic or Title with any Roman; to use heavy job letter or characters with different faces on the same job, as in railroad work; to have but one set of figures in German offices where Roman is also used; that but one lot of leaders is required for each body, etc. Not only are all faces of each

12345 abcdefghijklmnopqrstuvwxyz 67890
ABCDEFGHIJKLMNOPQRSTUVWXYZ&
12345 ABCDEFGHIJKLMNOPQRS 67890

One of the most important changes is our new lining system. A glance at the specimen sheets issued during

BOOK  FACES

THE STANDARD LINING SYSTEM

One of the most important changes is our lining system. A glance at specimen sheets issued during the last few years shows that the constant demand for something of this kind has led to ever recurrent attempts to solve the problem; but these efforts have been sporadic and inconsistent; failure to take into account all the conditions has rendered the results very unsatisfactory. Our type, including Romans Italics, Titles, Antiques, Gothics, and all other job faces, is "Standard" Line, and therefore all faces of one body line with one another. The advantages of this system are so many that it were difficult to enumerate all of them. Some which may be mentioned are: That it is now possible to line any Italic or Title with any Roman; to use heavy job letter or characters

abcdefghijklmnopqrstuvwxyz
ABCDEFGHIJKLMNOPQRSTUVWXYZ&
12345 ABCDEFGHIJKLMNOPQRSTU 67890

One of the most important changes is our new lining system. A glance at the specimen sheets

STANDARD LINING SYSTEM

One of the most important changes is our lining system. A glance at specimen sheets issued during the past few years shows that a constantly growing demand for something of this kind has led to ever recurrent efforts to solve this perplexing problem; but these attempts have been sporadic and inconsistent, and failure to take into account all the conditions has rendered results unsatisfactory. All our type, including Romans, Italics, and all other job faces, is "Standard" Line, and faces on one body line with one another perfectly. The advantages of this system are many, and it would be very difficult to enumerate all of them. Among these are

abcdefghijklmnopqrstuvwxyz
ABCDEFGHIJKLMNOPQRSTUVWXY
12345 ABCDEFGHIJKLMNOPQRST 67890

One of the most important features is our lining system. In a glance at the specimen

INLAND TYPE FOUNDRY

ST. LOUIS, MO., U. S. A.

BOOK

STANDARD LINE · IIF · HANOVER PA · INLAND TYPE

FACES

5-POINT ROMAN NO. 25

THE STANDARD LINING SYSTEM

One of the most important changes is our lining system. One glance at the specimen sheets issued during the past few years will show that the constant demand for something of this kind has led to ever recurrent attempts to solve the problem; these efforts, however, have been sporadic and inconsistent, failure to take into account all the conditions having rendered the results unsatisfactory. All our type is cast "Standard" Line, including Romans and Italics, Titles, Antiques, Gothics, and all other job faces; therefore, all faces of one body line accurately with one another. The advantages of this new system are so many that it would be difficult to enumerate all of them. Among those which may be mentioned are: That it is now possible to line any Italic or Title with any Roman; to use heavy job letter, figures or other characters with different faces on the same job, as in railroad work, etc.; to have but one set of figures need be purchased for each body, etc. Not only are all the faces of each body on the same line, but faces of different bodies justify in line with one another by use of 2-Point and 1-Point leads, the latter being necessary only on the smaller bodies. As the spaces of all bodies are point set, fractions or multiples of points, they can be used for this justification as well, thereby enabling the compositor to use the caps of the next smaller size of the series for small faces line with the standard leaders, but will line with 2-Point single or dotted rule perfectly by use of 2-Point or 1-Point leads and standard quads. In spite of the general use of leaders it is often necessary to use rule, and this innovation, which enables the printer to have accurate lining without the use of paper or cardboard, will be highly appreciated. The f's and j's are in all cases, except Italics and Scripts, made non-kerning, doing away not only with the useless expense of the old f-combinations, but securing absolute immunity from breakage after the type is in

12345 abcdefghijklmnopqrstuvwxyz 67890
ABCDEFGHIJKLMNOPQRSTUVWXYZ&
12345 ABCDEFGHIJKLMNOPQRSTUVWXYZ& 67890

One of the most important changes is our lining system. A glance at the specimen sheets issued during the past few years will show that

INLAND TYPE FOUNDRY

5½-POINT ROMAN NO. 25

THE STANDARD LINING SYSTEM

One of the most important changes is our new system of lining. A glance at the specimen sheets issued during the past few years shows that a constant demand for something of this kind has led to ever recurrent attempts to solve the problem. These efforts, however, have been sporadic and inconsistent, failure to take into account all the conditions having rendered the results unsatisfactory. All our type is cast on "Standard" Line, including Romans, Italics, and all job faces; therefore, all faces on the same body line with one another. The advantages of this system are so many that it would be difficult to enumerate all of them. Among those which may be mentioned are: That it is now possible to line any Italic or Title with any Roman; to have but one set of figures in German offices where Roman is also being used; to use heavy job letter, figures or characters with different faces on the same job, as in railroad work; that but one lot of leaders is necessary for each body, etc. Not only are all the faces of each body on the same line, but all faces of different bodies justify in line with one another by the use of 2-Point or 1-Point leads, the latter size lead being necessary only on the smaller bodies. The spaces of every body being point set, fractions or multiples of points, they can be used for this justification as well. This feature is of the greatest importance in job work, and by enabling the compositor to use the caps of the next smaller size of the series for small caps results in saving much material. Not only will all faces line with the standard leaders, but they will line perfectly with 2-Point single or dotted rule by the use of 2-Point or 1-Point leads and standard quads. In spite of the general use of leaders, it is often necessary to use rule, and this innovation, which enables the printer to have

12345 abcdefghijklmnopqrstuvwxyz 67890
ABCDEFGHIJKLMNOPQRSTUVWXYZ&
12345 ABCDEFGHIJKLMNOPQRSTUVWXYZ& 67890

One of the most important changes is our lining system. A glance at the specimen sheets issued during recent years will

ST. LOUIS, MO., U. S. A.

BOOK

STANDARD LINE · ITF · INLAND TYPE FOUNDRY

FACES

6-POINT ROMAN NO. 23

THE STANDARD LINING SYSTEM

One of the most important changes is our new lining system. A glance at the specimen sheets issued in the past few years will show that a constant demand for something of this kind has led to ever recurrent efforts to solve the problem; but these attempts have been sporadic and inconsistent, and the failure to take into account all the conditions has caused the results to be unsatisfactory. All our type, including Romans and Italics, Titles, Antiques, Gothics, and all job faces, is cast "Standard" Line; therefore all faces on the same body line with one another. The advantages of this system are so many that it would be quite difficult to enumerate all of them. Amongst those which may be mentioned are: It is now possible to line any Italic or Title with any Roman; to use heavy job letter, figures or characters with different faces on the same job, as in railroad work; in German offices, where Roman is also used, to have but one set of figures; that but one lot of leaders need be bought for each body, etc. Not only are all the faces of each body on the same line, but faces of different bodies will justify in line with one another by the use of 2-Point and 1-Point leads, the latter being necessary only on the smaller bodies. As the spaces of all bodies are on point set, fractions or multiples of points, they can be used for justification of this nature as well. This feature is of the greatest importance in job work, and by enabling compositors to use the caps of the next smaller size of the series for

12345 abcdefghijklmnopqrstuvwxyz 67890

ABCDEFGHIJKLMNOPQRSTUVWXYZ&

12345 ABCDEFGHIJKLMNOPQRSTUVWXYZ& 67890

One of the most important changes is our new system of lining. A glance at the specimen sheets issued during

INLAND TYPE FOUNDRY

6-POINT ROMAN NO. 26

THE STANDARD LINING SYSTEM

One of the most important changes is our lining system. A glance at the specimen sheets issued during the past few years will show that the constant demand for something of this kind has led to ever recurrent attempts to solve the problem; but these efforts have been sporadic and inconsistent, and the failure to take into account all the conditions having rendered the results unsatisfactory. All our type is cast on the "Standard" Line system, including Romans, Italics, Titles, Antiques, Gothics, and all other job faces; therefore all faces of one body line with one another. The advantages of this system are so many that it would be difficult to enumerate them all. Among those which may be mentioned are: It is now possible to line any Italic or Title with any Roman; to use heavy job letter, figures or other characters with different faces on the same job, as in railroad work; to have but one set of figures in German offices where Roman is also used; that but one lot of leaders for each body need be purchased, etc. Not only are all the faces of each body on the same line, but faces of different bodies justify in line with one another perfectly by use of 2-Point or 1-Point leads, the latter being necessary only on the smaller bodies. As the spaces of all bodies are point set, fractions or multiples of points, they can be used for this justification as well. This feature is of the greatest importance in job work, and by enabling compositors to use the caps of the next smaller size of the series as small caps results in the saving of much material. Not only will all faces line with the standard leaders, but will line with 2-Point single or dotted rule perfectly by the use of 2-Point or 1-Point leads and standard quads. In spite of the general use of leaders

12345 abcdefghijklmnopqrstuvwxyz 67890

ABCDEFGHIJKLMNOPQRSTUVWXYZ&

12345 ABCDEFGHIJKLMNOPQRSTUVWXYZ& 67890

One of the most important changes is our new lining system. A glance at the specimen sheets issued during the past few years shows

St. Louis, Mo., U. S. A.

BOOK FACES

6-POINT ROMAN NO. 27

THE STANDARD LINING SYSTEM

One of the most important changes is our new system of lining. A glance at the specimen books issued during the past few years will show that the constant demand for something of this kind has led to ever recurrent attempts to solve the problem; but these efforts have been sporadic and inconsistent, and failure to take into account all the conditions has rendered the results unsatisfactory. All our type, including Romans and Italics, Titles, Antiques and Gothics, as well as all other job faces, is cast on the improved "Standard" Line, and therefore all faces of one body will line with one another. The advantages of this system are so many that it would be difficult to enumerate all of them. Among those which may be here mentioned are: That it is now possible to line any Italic or Title with any Roman; to use heavy job letter, figures or characters with varying faces on the same body, as in railroad work; to have but one set of figures in German offices where Roman is also used; that but one lot of leaders need be purchased for each body, etc. Not only are all the faces of each body on the same line, but faces of different bodies will justify in line with one another readily by the use of 2-Point and 1-Point leads, the latter size being necessary only on the smaller bodies. As the spaces of all bodies are cast on point set, fractions and multiples of points, they can be used for this justification as well. This feature is of the greatest importance in job work, and by enabling the compositor to use the caps of the next smaller size of the series for small caps results in the saving of much material. Not

12345 abcdefghijklmnopqrstuvwxyz 67890
ABCDEFGHIJKLMNOPQRSTUVWXYZ&
12345 ABCDEFGHIJKLMNOPQRSTUVWXYZ& 67890

One of the most important changes is our lining system. A glance at the specimen sheets issued during the past few years

7-POINT ROMAN NO. 27

THE STANDARD LINING SYSTEM

One of the most important changes is our system of lining. A glance at the specimen sheets issued in the past few years will show that a constant demand for something of this kind has led to ever recurrent attempts to solve the problem; but these efforts have been sporadic and inconsistent, the failure to take into account all the conditions having rendered the results very unsatisfactory. All our type, including Romans and Italics, Titles, Antiques, Gothics, and all other job faces, is cast "Standard" Line, because of which all the faces of one body will line with one another. The advantages of this system are so many that it would be difficult to enumerate them all; but among those that may be mentioned are: That it is now possible to line every Italic or Title with every Roman; to use heavy job letter, figures or characters on the same job, as in railroad work; that but one set of leaders need be bought for each body, etc. Not only are all faces of each body on the same line, but faces of different bodies will justify in line with one another accurately by the use of 2-Point or 1-Point leads, the latter being necessary only on the smaller bodies. As the spaces of all bodies are cast on point sets, fractions or multiples of points, these can easily

12345 abcdefghijklmnopqrstuvwxyz 67890
ABCDEFGHIJKLMNOPQRSTUVWXYZ&
12345 ABCDEFGHIJKLMNOPQRSTUVWXYZ& 67890

One of the most important changes is our new lining system. A glance at the specimen sheets issued during

ST. LOUIS, MO., U. S. A.

☞ The No. 27 Series is Cast on Point and Half-Point Sets

INLAND TYPE FOUNDRY

BOOK

FACES

8-POINT ROMAN NO. 27

THE STANDARD LINING SYSTEM

One of the important changes is our system of lining. A glance at the specimen sheets that were issued during the past few years will show that the constant demand for something of this kind has led to ever recurrent attempts to solve the problem; but these efforts have been sporadic and inconsistent, failure to take into account all the conditions having rendered the results unsatisfactory. All our type, including Romans and Italics, Titles, Antiques, Gothics, and all other job faces, is cast on "Standard" Line, and therefore all faces of one body will line with one another. The advantages of this system are so many that it would be difficult to enumerate all of them. Among those that can be mentioned are: That it is now possible to line any Italic or Title with any Roman; to use heavy job letter and figures or characters with different faces on the same job, as in railroad work; to have but one set of figures in every German office where Roman is also used; that but one set of leaders

12345 abcdefghijklmnopqrstuvwxyz 67890
ABCDEFGHIJKLMNOPQRSTUVWXYZ&
12345 ABCDEFGHIJKLMNOPQRSTUVWXYZ& 67890

One of the most important changes is our system of lining. A glance at the specimen sheets is issued

INLAND TYPE FOUNDRY

PRACTICAL DEMONSTRATION OF

OUR NEW LINING SYSTEM

This paragraph, set in a number of our 8-Point faces, is a practical showing of our STANDARD LINE system. It illustrates how our standard line Half-Title.....our standard line Condensed No. 1...........our standard line Antique No. 1......our standard line *Gothic Italic*..........our standard line Latin...... our standard line **Gothic No. 1**...........our standard line Extended Old Style......our standard line Woodward.....our standard line Tudor Black, all line accurately with our standard line Roman No. 23 (in which this paragraph is set), *and its Italic;* how these also line with our standard line newspaper Roman No. 22 *and its Italic,* with our standard line Old Style No. 9 *and its Italic,* with our standard line French Old Style No. 8 *and its Italic;* and also how our standard line Round-Dot Leaders......our Fine-Dot Leaders..........line with these various faces. To this we also append sample words illustrating combinations of caps of 8-Point and 6-Point sizes, as caps and small caps: HALF-TITLE, ANTIQUE NO. 1, WOODWARD SERIES, LATIN SERIES, GOTHIC NO. 1, *GOTHIC ITALIC NO. 1;* all of which are justified in line with 1-Point leads.

BOOK

FACES

6-POINT FRENCH OLD STYLE NO. 8

7-POINT FRENCH OLD STYLE NO. 8

THE SUPERIOR STANDARD LINING SYSTEM

One of the most important changes is our new lining system. A glance at the specimen sheets issued during the past few years shows that a constantly increasing demand for something of this kind has led to ever recurrent attempts to solve the problem; but these efforts have been sporadic and inconsistent, and failure to take into account all the conditions has rendered results unsatisfactory. All our type is cast "Standard" Line, including Romans, Italics, Titles, Antiques and all other job faces, and therefore all faces of one body line with one another. The advantages of this system are many, and it would be difficult to enumerate all of them. Amongst those which could be mentioned are: That it is now possible to line any Italic or Title with any Roman; to use heavy job letter, figures or characters with other and different faces on the same job, as in railroad work; to have but one set of leaders need be purchased for each body, etc., etc. Not only are all the faces of each body on the same line, but the faces of different bodies justify in line with one another by using 2-Point or 1-Point leads, the latter being necessary only on smaller bodies. As all the spaces of all type bodies are point set, fractions or multiples of points, they can be used for this justification as well. This feature is of the greatest importance in job work, and by enabling the compositor to use the caps of the next smaller size of the series for small caps results in the saving of much material. Not only will all faces line with standard leaders, but the line has been placed in such position on the body that in every case the face will line with 2-Point rule by the use of 2-Point or 1-Point leads. In spite of the general use of leaders it is often necessary to use rule, and this innovation, which enables the compositor to have accurate line without the use of paper

abcdefghijklmnopqrstuvwxyz

ABCDEFGHIJKLMNOPQRSTUVWXYZ&

12345 ABCDEFGHIJKLMNOP 67890

One of the most important changes is our lining system. A glance at the specimen sheets issued during the past few years will show that a

INLAND TYPE FOUNDRY

SUPERIOR STANDARD LINING SYSTEM

One of the most important changes is our new system of lining. A glance at specimen sheets issued during the past few years shows that a constant demand for something of this kind has led to ever recurrent attempts to solve the vexing problem; but these efforts have been sporadic and inconsistent, and the failure to take into account all the conditions has rendered the results unsatisfactory. All our type, including Romans, Italics, and all job faces, is cast on "Standard" Line; therefore all faces of one body line with one another. The advantages of this system are many, and it would be difficult to enumerate them all. But among those which could be mentioned are: That it is now possible to line any Italic or Title with any Roman; to use heavy job letter, figures or characters with different faces on the same job, as in railroad work; to have but one set of figures in German offices where Roman is also used; that but one lot of leaders need be purchased for each body in the office, etc. Not only are all the faces of each body on the same line, but faces of different bodies justify in line with one another by the use of 2-Point or 1-Point leads, the latter being necessary only on the smaller bodies. As all the spaces of every type body are point set, fractions or multiples of points, they can be used for this justification as well. This feature is of the greatest importance in job

abcdefghijklmnopqrstuvwxyz

ABCDEFGHIJKLMNOPQRSTUVWXYZ&

12345 ABCDEFGHIJKLMNOP 67890

One of the most important changes is the new lining system. A glance at the specimen sheets which have

ST. LOUIS, MO., U. S. A.

BOOK ✦ FACES

8-POINT FRENCH OLD STYLE No. 6

SUPERIOR STANDARD LINING SYSTEM

One of the most important changes is our new lining system. A glance at the specimen sheets issued during the past few years will show that the increasing demand for something of this description has led to ever recurrent attempts to solve the problem; but these attempts have been sporadic and inconsistent, and failure to take into account all the conditions has rendered results anything but satisfactory. All our type, including Romans, Italics and all job faces, is cast on "Standard" Line; therefore all faces of one body line with one another. The advantages of this system are many, and it would be difficult to here enumerate all of them. Amongst others which might be mentioned are: That it is now possible to line any Italic or Title with any Roman; to use heavy job letter, figures or characters with different faces on the same job, as in railroad work; to have but one set of figures in German offices where Roman is also used; but one lot of leaders need be purchased for each body, etc., etc. Not only are all the faces of each body on the same line, but faces of different bodies justify in line with one another by using

abcdefghijklmnopqrstuvwxyz
ABCDEFGHIJKLMNOPQRSTUVWXYZ&
12345 ABCDEFGHIJKLMNOP 67890

One of the most important changes is the improved lining system. A glance at the specimen sheets issued during the

INLAND TYPE FOUNDRY

German, French, Spanish and Swedish
Accents are made for these faces

9-POINT FRENCH OLD STYLE No. 6

SUPERIOR STANDARD LINING SYSTEM

One of the most important changes is our system of lining. A glance at the specimen sheets issued during the past few years will show that the constant demand for something of this nature has led to ever recurrent attempts to solve the problem; but these efforts have been sporadic and inconsistent, and the failure to take into account all the conditions has rendered the results anything but satisfactory. All our type, including the Romans, Italics, Titles, Antiques, Gothics, and all other job faces, is cast on "Standard" Line; hence all faces of one body line with one another. The advantages of this system are so many that it would be difficult to enumerate all of them. But among those which may be mentioned are: That it is now possible to accurately line any Italic or Title with any Roman; to use heavy job letter, figures or characters with different faces on the same job, as in railroad work; one kind of leaders only need be purchased for each body used in an office

abcdefghijklmnopqrstuvwxyz
ABCDEFGHIJKLMNOPQRSTUVWXYZ&
12345 ABCDEFGHIJKLMNOP 67890

One of the most important changes is our new system of lining. A glance at the specimen sheets issued during

St. Louis, Mo., U. S. A.

German, French, Spanish and Swedish
Accents are made for these faces

10-POINT FRENCH OLD STYLE No. 8

12-POINT FRENCH OLD STYLE No. 8

STANDARD LINING SYSTEM

One of the most important changes is our new lining system. A glance at the specimen sheets issued during the past few years shows that the constant demand for something of this character has led to ever recurrent attempts to solve the problem; but these efforts have been sporadic and inconsistent, and failure to take into account all the conditions has rendered the results unsatisfactory. All our types are cast on "Standard" Line, including Romans, Italics and all job faces, consequently all faces on one body line with one another. The advantages of this system are so many that it would be difficult to enumerate all of them. It is now possible to line any Italic or Title with Roman

abcdefghijklmnopqrstuvwxyz
ABCDEFGHIJKLMNOPQRSTUVWXYZ&
12345 ABCDEFGHIJKLMNOP 67890

One of the most important changes is our new lining system. A glance at the specimen sheets

INLAND TYPE FOUNDRY

STANDARD LINING SYSTEM

One of the most important changes is our lining system. A glance at the specimen sheets issued during the last few years will show that the constant demand for something of this kind has led to ever recurrent attempts to solve the problem; these efforts have been sporadic and inconsistent, and failure to take into account all the conditions has rendered the results anything but satisfactory. All our type, including Romans, Italics, Titles, Gothics, and all job faces, is cast on "Standard" Line

abcdefghijklmnopqrstuvwxyz
ABCDEFGHIJKLMNOPQRSTUVWX
12345 ABCDEFGHIJKLMNOP 67890

One of the most important changes is our lining system. In glancing at the

ST. LOUIS, MO., U. S. A.

43

BOOK FACES

6-POINT OLD STYLE NO. 9

THE STANDARD LINING SYSTEM

One of the most important changes is our lining system. A glance at the specimen sheets issued during the past few years will show that the constant demand for something of this kind has led to ever recurrent attempts to solve the problem, but these efforts have been sporadic and inconsistent, and failure to take into account all the conditions has rendered the results unsatisfactory. All our type is cast "Standard" Line, including Romans, Italics, Titles, Antiques, Gothics, and all other job faces, therefore all faces of one body line with one another. The advantages of this new system are so manifold that it would be difficult to enumerate all of them. Amongst those which can be mentioned are: That it is now possible to line any Italic or Title with any Roman; to use heavy job letter, characters or figures with different faces on same job, as railroad work; to have but one set of figures in German offices where Roman is also used; that but one lot of leaders need be purchased for each body, etc., etc. Not only are all the faces of each body on the same line, but faces on different bodies justify in line with one another by the use of 2-Point or 1-Point leads, the latter size being necessary only on the smaller bodies. As the spaces of all bodies are point set, fractions or multiples of points, they can be used for this justification as well. This feature is of the greatest importance in job work, and by enabling compositors to use the caps of the next smaller size of the series for small caps, results in the saving of much material. Not only will all faces line with standard leaders, but the line has been placed in such position on the body that in every case the face will line with 2-Point rule by the use of 2-Point or 1-Point leads. In spite

abcdefghijklmnopqrstuvwxyz
ABCDEFGHIJKLMNOPQRSTUVWXYZ&
12345 ABCDEFGHIJKLMNOPQRS 67890

One of the most important changes is our lining system. A glance at the specimen sheets issued during the past years will

7-POINT OLD STYLE NO. 9

THE STANDARD LINING SYSTEM

One of the important changes is our lining system. A glance at the specimen sheets issued during the past few years will show that the constant demand for something of this kind has led to ever recurrent attempt to solve this vexing problem, but these efforts have been sporadic and inconsistent, and the failure to take into account all the conditions has rendered the results unsatisfactory. All our type, including Romans, Italics, Titles, Antiques and Gothics, and all other job faces, is on "Standard" Line, and therefore all faces of one body line with one another. The advantages of this system are so many that it would be difficult to enumerate them all. Among those which may be mentioned are: That it is now possible to line any Italic or Title with any Roman; to use heavy job letter, figures or characters with different faces on the same job, as in railroad work; to have but one set of figures in German offices where Roman is also used; that it is unnecessary to purchase more than one lot of leaders for each body in the office, etc. Not only are all the faces of every body on the same line, but faces of different bodies justify readily in line with one another by the use of 2-Point and 1-Point leads, the latter size being necessary only on the smaller bodies. As the spaces of all bodies are point set, fractions and multiples of points, they can be used for justification

abcdefghijklmnopqrstuvwxyz
ABCDEFGHIJKLMNOPQRSTUVWXYZ&
12345 ABCDEFGHIJKLMNOPQRS 67890

One of the most important changes is the system of lining. A glance at the specimen sheets printed in the

BOOK  FACES

9-POINT OLD STYLE NO. 9

THE STANDARD LINING SYSTEM

One of the most decided changes is our lining system. A glance at the specimen sheets issued during the past few years shows a constant and growing demand for something of this kind, one which has led to ever recurrent attempts to solve the problem; these efforts, however, have been sporadic and inconsistent, and failure to take into account all the conditions has rendered results unsatisfactory. All our type, including Romans and Italics, Titles, Antiques, Gothics, and all job faces, is cast on "Standard" Line, consequently all faces of one body line with one another. The advantages of this system are so many that it would be difficult to enumerate all of them, but among the advantages it is now possible to line any Italic or Title with any Roman; to use heavy job letter, figures, etc., with different faces on the same line, as in railroad work; to have but one

abcdefghijklmnopqrstuvwxyz

ABCDEFGHIJKLMNOPQRSTUVWXYZ&

12345 ABCDEFGHIJKLMNOPQR 67890

One of the most important changes is our lining system. A glance at the specimen sheets issued now

St. Louis, Mo. U. S. A.

8-POINT OLD STYLE NO. 9

THE STANDARD LINING SYSTEM

One of the most important changes is our lining system. A glance at specimen sheets issued during the past few years shows that the constant demand for something of this kind has led to ever recurrent attempts to solve the problem; but these efforts have been sporadic and inconsistent, and failure to take into account all the conditions has rendered results unsatisfactory. All our type is cast "Standard" line including Romans, Italics, Titles, Antiques, Gothics and all job faces, and therefore all faces of one body line with one another. The advantages of this system are so many that it would be difficult to enumerate all of them. Amongst those which can be mentioned are: That it is now possible to line any Italic or Title with any Roman; to use heavy job letter, figures or characters with different faces on the same job, as for railroad work; to have but one set of figures in German offices where Roman is also used; that but one set of leaders need be purchased for each body etc., etc. Not only are all faces of each body on the same line, but faces of different bodies justify in line

abcdefghijklmnopqrstuvwxyz

ABCDEFGHIJKLMNOPQRSTUVWXYZ&

12345 ABCDEFGHIJKLMNOPQRS 67890

One of the most important changes is our lining system. A glance at specimen sheets issued during

INLAND TYPE FOUNDRY

BOOK  FACES

German, French, Spanish and Swedish
Accents are made for these faces

10-POINT OLD STYLE NO. 9

THE STANDARD LINING SYSTEM

One of the most important changes is our lining system. A glance at specimen sheets issued during the past few years shows that a constant demand for something of this kind has led to ever recurrent attempts to solve the problem; but these efforts have been sporadic and inconsistent, failure to take into account all the conditions having rendered the results unsatisfactory. All our type is cast on the improved "Standard" Line, and all the faces of one body line with one another. Among the advantages of this system which may here be mentioned are: It is now possible to line any Italic or Title with any Roman; to use heavy job letter, figures, etc., with different faces on

abcdefghijklmnopqrstuvwxyz
ABCDEFGHIJKLMNOPQRSTUVWXYZ&
12345 ABCDEFGHIJKLMNOPQR 67890

One of the important changes is our lining system. A glance at the specimen sheets issued

11-POINT OLD STYLE NO. 9

STANDARD LINING SYSTEM

One of the most important changes is our lining system. A glance at specimen sheets issued during past few years shows that a constant demand for something of the kind has led to ever recurrent attempts to solve the problem. These efforts have been sporadic and inconsistent, failure to take into account all the conditions having rendered the results unsatisfactory. Our type, including Romans, Italics, Titles and all job faces, is cast on "Standard" Line, all faces on every body lining together. The advantages of this system are many, and it is difficult to enumerate all of them. But

abcdefghijklmnopqrstuvwxyz
ABCDEFGHIJKLMNOPQRSTUVWX
12345 ABCDEFGHIJKLMNOPQR 67890

One of the most important changes is our lining system. A glance at the specimens

ST. LOUIS, MO., U. S. A.

BOOK

FACES

STANDARD LINING SYSTEM

One of the most important changes is our lining system. A glance at the specimen sheets issued during the past few years will show that the persistent demand for something of this kind has led to ever recurrent attempts to solve the problem. These efforts have been sporadic and inconsistent, and failure to take into account all the conditions has rendered the results unsatisfactory. All our type is cast on "Standard" Line, and faces of one body line with one another

abcdefghijklmnopqrstuvwxyz
ABCDEFGHIJKLMNOPQRSTUV
12345 ABCDEFGHIJKLMNOP 67890

One of the most important changes is our lining system. A glance at specimen

INLAND TYPE FOUNDRY

SYSTEM OF LINING

One of the most important changes is our new system of lining. A glance through the specimen books recently issued shows that a constant demand for something of this kind has led to ever recurrent attempts to solve the vexing problem; but all these efforts have been most inconsistent

abcdefghijklmnopqrstuvwxz
ABCDEFGHIJKLMNOP
123 QRSTUVWXYZ 456

ST. LOUIS, MO., U. S. A.

BOOK

FACES

6-Point Old Style No. 10

THE STANDARD LINING SYSTEM

One of the most important changes is our new system of lining. A glance at the specimen sheets issued during the past few years shows that a constant demand for something of this kind has led to ever recurrent attempts to solve the vexing problem; but these efforts have been sporadic and inconsistent, and the failure to take into account all the conditions has caused the results to be unsatisfactory. All our type, including Romans, Italics, Antiques, Gothics and all job faces, is cast on "Standard" Line; therefore all our faces of one body line with one another. The advantages of this system are so many that it would be quite difficult to enumerate all of them. We could mention among others that: It is now possible to line any Italic or Title with any Roman: to use heavy job letter, figures or characters with different faces on the same job, as in railroad work: to have but one set of figures in German offices where Roman is also used; that but one lot of leaders need be purchased for each body, etc. Not only are all the faces of each body on the same line, but faces on different bodies justify in line with one another readily by the use of 2-Point and 1-Point leads, the latter size being necessary only on the smaller bodies. As the spaces of all bodies are point set, fractions or multiples of points, they can be used for this justification as well. This feature is of the greatest importance in job work, and by enabling compositors to use the caps of the next smaller size of the series for small caps, results in the

abcdefghijklmnopqrstuvwxyz

ABCDEFGHIJKLMNOPQRSTUVWXYZ&
12345 ABCDEFGHIJKLMNOPQRST 67890

One of the most important changes is our system of lining. A glance at the specimen sheets issued during

8-Point Old Style No. 10

THE STANDARD LINING SYSTEM

One of the most important changes is our new lining system. A glance at the specimen sheets issued during the past few years will show that the constant demand for something of this kind has led to ever recurrent attempts to solve the problem; but these efforts have been sporadic and inconsistent, failure to take into account all conditions having rendered the results anything but satisfactory. All our types being cast on the new "Standard" Line, including Romans, Italics and all job faces, all faces on one body line with one another. The advantages of this system are so many that it would be difficult to enumerate all of them. We could mention among others that: It is now possible to line any Italic or Title with any Roman; to use heavy job letter, figures or characters with different faces on same job, as in railroad work; to have but one set of figures in German offices where Roman is also used; but one lot of leaders needed for each body, etc. Not

abcdefghijklmnopqrstuvwxyz

ABCDEFGHIJKLMNOPQRSTUVWXYZ&
12345 ABCDEFGHIJKLMNOPQRST 67890

One of the most important changes is our lining system. A glance at the specimen sheets issued now

German, French, Spanish and Swedish
Accents are made for these faces

9-POINT OLD STYLE No. 10

STANDARD LINING SYSTEM

One of the most important changes is our new lining system. A glance at the specimen sheets issued during the past few years will show that the constant demand for something of this kind has led to ever recurrent attempts to solve the problem; but these efforts have been sporadic and inconsistent, failure to take into account all conditions having rendered the results anything but satisfactory. All our types being cast on the new "Standard" Line, including Romans, Italics and all job faces, all faces on one body line with one another. The advantages of this system are so many that it would be difficult to enumerate all of them. We could mention among others that: It is now possible to line all Italic or Title with all Roman; to use heavy job letter, figures or characters with different faces on all jobs, as

abcdefghijklmnopqrstuvwxyz.
ABCDEFGHIJKLMNOPQRSTUVWXYZ
12345 ABCDEFGHIJKLMNOPQRST 67890

One of the most important changes is our new lining system. A glance at the specimen sheets

INLAND TYPE FOUNDRY

10-POINT OLD STYLE No. 10

STANDARD LINING SYSTEM

One of the most important changes is our lining system. A glance at the specimen sheets issued during the past few years will show that a constant demand for something of this description has led to ever recurrent attempts to solve the problem; but these efforts were sporadic and inconsistent, and failure to take into account all conditions has rendered results unsatisfactory. Our types are cast "Standard" Line, including Romans, Italics, Titles, Gothics, Antiques and all other job faces; therefore, all faces on one body line with one another. The advantages of this system are so many that it would be difficult to here enumerate all

abcdefghijklmnopqrstuvwxyz
ABCDEFGHIJKLMNOPQRSTUVWXY
12345 ABCDEFGHIJKLMNOPQRS 67890

One of the most important changes is our lining system. A glance at the specimens is

ST. LOUIS, MO., U. S. A.

49

INVITATION SCRIPT

20a 8A, \$7.50   24-POINT INVITATION SCRIPT   L. C. \$3.75; C. \$3.75

*Our Script faces are cast on systematic
lines and unit sets, like all the other type
made by the Inland Type Foundry,
hence in their usefulness are far superior
to the Scripts of the other foundries: $84*

CARD FONTS, 10a 4A, \$4.25 — L. C. \$2.15; C. \$2.10

25a 9A, \$6.00   18-POINT INVITATION SCRIPT   L. C. \$3.00; C. \$3.00

*In addition to the adoption of Standard Line
and Unit Set, in the casting of these Scripts, other
desirable features are worthy of special notice, such
as joined apostrophe and s, s. the logotype &c. double
beginning and ending strokes, one with hyphen
added, which add considerably to its appearance. 3*

CARD FONTS, 12a 4A, \$3.25 — L. C. \$1.65; C. \$1.60

35a 10A, \$5.00   12-POINT INVITATION SCRIPT   L. C. \$2.70; C. \$2.30

*Another excellent feature in our Scripts may be noted in the Figures
which, instead of straggling above and below the line, are cut uniform
in size and cast regular in line, the same as our popular lining figures
in Old Style fonts. In the following 4 pages we exhibit the use of the
Invitation Script, on samples of society work showing the latest usages
of fashion, which now stamps with its high approval this style of letter.*

CARD FONTS, 18a 5A, \$2.80 — L, C, \$1.60; C. \$1.20

SPACES AND QUADS ARE INCLUDED IN ALL SCRIPT FONTS.

Mr. and Mrs. Henry Jordan
request the honor of your presence
at the marriage of their daughter

Francesca

to

Mr. August Rankin,

on Monday evening November the eighth,
eighteen hundred and ninety-seven,
at half past eight o'clock
3571 Clinton Place,
St. Louis

At Home
after December the first,
3204 Robinsons Boulevard

Miss Ophelia Capulet,

At Home

Tuesday, evening, February, fifteenth,

from, eight, to, twelve, o'clock,

Pine Ridge,

Hills' Point, Missouri.

Dancing: *R. S. V. P.*

Reception

from, half, after, eight, to

half, after, eleven, o'clock

Vista House:

Mrs. T. C. Bernheim

announces the marriage of her daughter

Florence Isabell

to

Mr. Charles Lamonte,

on Wednesday, August the fourth,

eighteen hundred and ninety-seven,

Edwardsville, Illinois.

The sample on page 51 is the correct form for a Wedding Invitation. It is set in 24-Point, 18-Point and 12-Point INVITATION SCRIPT, and should be printed on the first page of a folio sheet; size of page as indicated. It should be folded once and put in an envelope.

If it is a church wedding, an Admission Card, size 2 by 3½ inches, bearing in the centre the line "Please present this card at the Church," printed in 18-Point Invitation Script, may be enclosed.

The hour of ceremony is often announced on a separate card, size 2 3-8 by 3 1-4 inches, the word "Ceremony" being set in 24-Point, and the line "at five o'clock" in 18-Point Invitation Script, immediately below it, separated by a 6-Point slug; printed in the centre of the card.

For those whose presence is desired at the reception after the ceremony the card at the bottom of page 52 is enclosed.

The Announcement, size and style as on this page, should be printed, folded and enveloped the same as a wedding invitation. With it should be enclosed an At Home Card, size and style of the one on page 54.

At Home

Thursdays in October,

3970 Cleveland Avenue,

Tyler Place.

Mr. Harvey B. Dickens.

Miss Frances S. Coulter.

Mrs. James U. Easton.

The card at the top of page 52 shows the correct size and form of an Invitation to a dance at a private residence.

The smaller cards on this page are the correct sizes for gentlemen, married and single ladies.

Fridays.   *4692 Wilmington Avenue.*

INLAND TYPE FOUNDRY   54   ST. LOUIS, MO., U. S. A.

20a 7A, $7.50   24-POINT COMMERCIAL SCRIPT   L. C. $3.75; C. $3.75

All these Script faces are cast on uniform line, enabling the compositer to justify 2-point rule, single or dotted, in position to line accurately in blank work, avoiding card-board 52

CARD FONTS, 10a 4A, $4.25 — L. C. $2.10; C. $2.15

25a 9A. $6.00   18-POINT COMMERCIAL SCRIPT   L. C. $3.00; C. $3.00

Regular Standard Line **Job Faces** *can be used in combination with our Scripts, in cases where special* **emphasis** *is considered desirable, the justification on a common line being readily accomplished by means of ordinary point system, leads and slugs.* 330

CARD FONTS, 12a 4A, $3.25 — L. C. $1.65; C. $1.60

36-POINT AND 48-POINT SIZES OF COMMERCIAL SCRIPT IN PREPARATION.

SPACES AND QUADS ARE INCLUDED IN ALL SCRIPT FONTS

ROYAL ITALIC

10a 6A. $3.50   24-POINT ROYAL ITALIC   L. C. $1.70; C. $1.80

HANDSOME DESIGN
Popular Attractions Found
Right Types Noticed 35

26a 10A, $3.00   14-POINT CALEDONIAN ITALIC   L. C. $1.75; C. $1.25

No. ....   *$* ....

Sixty Days after ....*we will pay the INLAND TYPE FOUNDRY, at St. Louis, in cash,* ....*Dollars, for invoice of Standard Line Type, Inland Art Ornaments, 144 feet of Borders and sixteen fonts of Labor-saving Rule.* .*
Prosperous Printers

28a 12A, $2.80   12-POINT CALEDONIAN ITALIC   L. C. $1.55; C. $1.25

TO POSTERITY: *Gondar, B. C. 56*
Two Thousand Years after ....
I promise to pay to some Type Foundry that makes Standard Line Type, 5,927.21 Dollars; if more time is required note will be made to suit; a complete catalog must be furnished not later than December ....*, A. D. 189* .....
No. ....   *Wise King Solomon*

34a 14A, $2.50   10-POINT CALEDONIAN ITALIC   L. C. $1.40; C. $1.10

JOYOUS GREETING!   *Age of Invention, 1895*
Be it known that on the ....*day of* ....*, 18* .....
appeared before us The Inland Type Foundry, inventors of Standard Line Type, and submitted proofs that no special leaders are required for their Caledonian Italic series; it lines with all their other faces on like bodies.
(Seal)   *Printing Fraternity*

............*HHn* ......*nHHHn* ....*nHll* ....

STANDARD LINE TYPE

Is the only material for the progressive printer, and its advantages are acknowledged everywhere. Both in finish and exactness our type is far superior to all others, will wear better and satisfy all requirements which arise in legal blank work as well as general 16

38a 14A, $2.50   10-POINT ITALIC No. 20   L. C. $1.40; C. $1.10

ALL OUR OLD STYLE FIGURES

Are uniform in height and line, and all small cap letters liable to be confused with the lower case characters have an extra nick near top of body. Old Style type has one nick more than Roman and our standard job faces are more uniform in face and in the finish, accuracy and durability are far superior to all others 47

44a 16A, $2.25   8-POINT ITALIC No. 20   L. C. $1.30; C. $0.95

ALL FACES OF ONE BODY LINE PERFECTLY

And faces of different bodies are made to line by use of 1-Point or 2-Point leads, without the use of card or paper. Throughout the old style series the figures are uniform in size and line, while retaining the old style design, thus adapting them to all uses. In all respects our type is the best, and being cast on point bodies will match those of all other type foundries in this respect 58

50a 18A, $2.00   6-POINT ITALIC No. 20   L. C. $1.20; C. $0.80

STANDARD LEADERS AND STANDARD LINE TYPE

Lining perfectly and labor-saving, thus adding to the profits while making pleasing effects in progressive and practical printing. In spite of the general use of leaders it is often necessary to use rule, and in this connection our standard line system will be appreciated, enabling the compositor to have accurate line without the use of cardboard, thus saving much time and labor. Not only will the adoption of our system save enormously in labor and produce better results, but as our material is more available for different classes of work there is an actual

............HHιιιι.......ιιΗΗHHιιιι......ιιHHH............

FRENCH O. S. ITALIC SERIES

Figures are Uniform in Size and Lining

36a 12A, $2.80   12-POINT FRENCH OLD STYLE ITALIC   L. C. $1.60; C. $1.20

OUR TYPE IS SUPERIOR TO ANY

These are excellent faces for circulars and being cast on our new system line with all job faces and Romans All faces on different bodies can be made to line with one another by the use of regular leads. Standard Line Leaders will line with all Romans and Old Styles 360

40a 16A, $2.50   10-POINT FRENCH OLD STYLE ITALIC   L. C. $1.45; C. $1.05

SUCCESSFUL, PRUDENT PRINTERS WHO

Used Standard Line Type will have nothing else. Our book and newspaper faces are now shown complete from 5-Point to 12-Point Printer will note that the Old Styles have lining figures throughout Our Leaders are also cast on point set and line with all faces. A glance at the specimen pages shown in this catalog will prove an

48a 18A, $2.25   8-POINT FRENCH OLD STYLE ITALIC   L. C. $1.30; C. $0.95

IT FILLS A LONG FELT WANT AMONG PRINTERS

Do not be imposed upon by the representations of other concerns who intimate that the advantages of our system are not as great as we claim. The great saving accomplished by Standard Line can be demonstrated by a practical test. We will gladly extend to you every facility for making such a test, and are willing to abide by the results. In finish and accuracy our type is superior to others 78

55a 20A, $2.00   6-POINT FRENCH OLD STYLE ITALIC   L. C. $1.20; C. $0.80

IT IS NOT THE NUMBER OF POUNDS OF TYPE YOU PURCHASE

Which counts, but the amount of material which is actually available. Look at the small amount of type you have set up in jobs and compare it with the large quantity which is lying idle in the cases. By adopting our system you can make the minimum quantity of type do the maximum amount of work, as it is available under any and all circumstances, while the old stuff can only be used for some particular purpose. Figure out the amount you pay yearly for labor. A saving of 5 or 10 per cent. will go a long ways toward purchasing the type actually used in your shop 492

...........HHHH......HHHHHHHH......HHHH

32a 14A, $2.80   12-POINT OLD STYLE ITALIC No. 9   L. C. $1.45; C. $1.35

ALL OLD STYLE FIGURES ARE

Uniform in size and line. In our system each figure not only matches its fellows in width, but is a multiple of one of the spaces. As this makes them multiples of points, they will justify with all other spaces, quads, leads, slugs and brass rules. Send for our catalog 68

38a 15A, $2.50   10-POINT OLD STYLE ITALIC No. 9   L. C. $1.40; C. $1.10

IN ALL THE OLD STYLES THE FIGURES

Above and below the line have been abandoned, and while they still retain their Old Style characteristics they have been made uniform in appearance, and their legibility increased. What has been said of the figures applies as well to the points. In all cases these will be found to justify with spaces and point 42

40a 15A, $2.25   8-POINT OLD STYLE ITALIC No. 9   L. C. $1.30; C. $0.95

IN CONNECTION WITH OUR LINING TYPE

But one set of Leaders is necessary for each body, since these Leaders will line with our Roman, Old Style, Job and patented faces with perfect results. Either the saving of original outlay or composition alone will go a long way towards making you a good profit in your composing room. The round-dot leaders are made on all bodies up to and including........18-Point, and the fine-dot..............up to 14-Point

45a 18A, $2.00   6-POINT OLD STYLE ITALIC No. 9   L. C. $1.15; C. $0.85

THE IMPROVED STANDARD LINING SYSTEM

One of the most important changes is our lining system. A glance at specimen sheets issued during the past few years will show that the constant demand for something of this kind has led to ever recurrent attempts to solve the problem; but these efforts have been sporadic and inconsistent, and failure to take into account all the conditions has rendered the results unsatisfactory. All our types are cast Standard Line, including Romans, Italics, Gothics, Latins, and all other Job faces, and line with one another 364

..................*HHH*n......n*HHHH*nn....n*HHH*..................

32a 14A, $2.80   12-POINT ROMAN NO. 20   L. C. $1.55; C. $1.25

OUR PERFECT SYSTEM OF LINING
STANDARD LINE LEADERS DESIRABLE
One of the most important changes is our lining system. A glance at the specimen sheets issued during the past few years shows that constant 258

40a 16A, $2.50   10-POINT ROMAN NO. 20   L. C. $1.40; C. $1.10

ROMANS, GOTHICS, ITALICS, TITLES, AND ALL
JOB FACES ON THE SAME BODY LINE PERFECTLY
One of the most important changes is our lining system. A glance at the specimen sheets issued during the past few years shows that a constant demand for something of this kind 360

45a 16A, $2.25   8-POINT ROMAN NO. 20   L. C. $1.30; C. $0.95

HANDSOMELY CUT FACES ON STANDARD LINE SYSTEM
EACH FACE WILL LINE PERFECTLY WITH EVERY OTHER
One of the most important changes is our lining system. A glance at the specimen sheets issued during the past few years will show that the constant demand for something of this nature has led to ever recurrent attempts to solve the problem; but these efforts have been sporadic $492

50a 20A, $2.00   6-POINT ROMAN NO. 20   L. C. $1.20; C. $0.80

FAVORITE OLD AS WELL AS ORIGINAL FACES ON IMPROVED LINE
SUPERIOR LABOR-SAVING FEATURES APPLIED TO PRINTING MATERIAL
One of the most important changes is our lining system. One glance at the specimen sheets issued during the past few years will show that the constant demand for something of this kind has led to ever recurring attempts to solve the problem; but these efforts have been sporadic and inconsistent, and the failure to take into account all the conditions 635

42a 16A, $2.00   5-POINT ROMAN NO. 20   L. C. $1.15; C. $0.85

STANDARD LINE LEADERS CAN BE USED FOR OTHER THAN ROMAN FACES
USEFULNESS OF LEADERS IS GREATLY AUGMENTED UNDER OUR SYSTEM
One of the most important changes is our lining system. A glance at the specimen sheets that were issued during the past few years shows that the constant demand for something of this kind has led to recurring attempts to solve the vexing problem; but these efforts have been sporadic and inconsistent, and the failure to take into account all the conditions has rendered the results very 480

SMALL CAP FONTS: 12-POINT, 9A, 50C.   10-POINT, 12A, 50C.   8-POINT, 14A, 45C.
6-POINT, 15A, 40C.   5-POINT, 10A, 40C. EXTRA.

HHHHHᴴᴴᴴHHHH

22a 12A, $3.20   16-POINT OLD STYLE NO. 9   L. C. $1.60; C. $1.60

POPULAR OLD STYLES
Improved with Standard Line Feature
Surpassing in Utility its Rivals 36

Cast also on 18-Point body to order.

34a 14A, $2.80   12-POINT OLD STYLE NO. 9   L. C. $1.55; C. $1.25

OUR FIGURES ARE UNIFORM IN SIZE
JUSTIFY WITH REGULAR SPACES OR QUADS
One of the most important changes is our system
of lining. A glance at the specimen sheets issued 950

40a 16A, $2.50   10-POINT OLD STYLE NO. 9   L. C. $1.40; C. $1.10

OUR TYPE IS BY FAR THE MOST DURABLE
GIVING SUPERIOR RESULTS IN ELECTROTYPING
One of the most important changes is our lining system. A
glance at the specimen sheets issued during the past few years
shows that the constant demand for something of this sort 486

45a 16A, $2.25   8-POINT OLD STYLE NO. 9   L. C. $1.30; C. $0.95

STANDARD LINE TYPE THE GREAT SAVER OF LABOR
TWENTIETH CENTURY TYPE SURE TO PLEASE PRINTERS
One of the most important changes is our lining system. A glance
at the specimen sheets issued during the past few years will show that a
constant demand for something of this nature has led to recurrent $372

50a 20A, $2.00   6-POINT OLD STYLE NO. 9   L. C. $1.20; C. $0.80

ALL OUR OLD STYLE FONTS ARE CAST WITH AN EXTRA NICK
READILY DISTINGUISHED FROM SAME SIZES OF MODERN FACES
One of the most important changes is our lining system. A glance at the specimen
sheets issued during the past few years shows that a constant demand for something
of this kind has led to ever recurrent attempts to solve the problem; these efforts 465

SMALL CAP FONTS: 12-POINT, 9A, 50C. 10-POINT, 12A, 50C.
8-POINT, 12A, 45C. 6-POINT, 15A, 40C. EXTRA.

LARGER SIZES OF THE OLD STYLE NO. 9 SERIES ARE IN PREPARATION.

HHHHHHIIIHHHHH

5a 4A, $7.25   48-POINT FRENCH OLD STYLE   L. C. $3.00; C. $4.25

MODERN
Ornate Face

7a 4A, $5.00   36-POINT FRENCH OLD STYLE   L. C. $2.50; C. $2.50

FURNISHED
First-class Slugs

9a 6A, $4.30   30-POINT FRENCH OLD STYLE   L. C. $2.00; C. $2.30

PROFIT-MAKING
Reduced to a Science

12a 6A, $3.50   24-POINT FRENCH OLD STYLE   L. C. $1.75; C. $1.75

NIMBLE PRINTERS
Produce Annual Results

15a 8A, $3.30   20-POINT FRENCH OLD STYLE   L. C. $1.65; C. $1.65

POPULAR OLD STYLE
Ornaments the Superior Work

26a 12A, $3.20   16-POINT FRENCH OLD STYLE   L. C. $1.70; C. $1.50

SEND FOR ESTIMATES NOW
This Believed a Reasonable Request 52

36a 15A, $2.80   12-POINT FRENCH OLD STYLE   L. C. $1.60; C. $1.20.

LINING FIGURES AND HANDSOME FACES
ARE DELIGHTS TO ARTISTIC PRINTERS GENERALLY
Fast Compositors Praise Standard Line 10

40a 16A, $2.50   10-POINT FRENCH OLD STYLE   L. C. $1.45; C. $1.05

BEST METAL COMPOSITION YET PRODUCED
PRACTICAL DEMONSTRATION WARRANTS THIS ASSERTION
Investigate the Merits of Standard Line Type 843

8-POINT FRENCH OLD STYLE	6-POINT FRENCH OLD STYLE
50a 18A, $2.25 L. C. $1.30; C. $0.95	55a 20A, $2.00 L. C. $1.20; C. $0.80
SUPERIOR FACILITIES	ORDERS PROMPTLY FILLED
ENABLING US TO PRODUCE BETTER	WITH POPULAR FACES ON STANDARD LINE
And More Durable Goods 23	Business Constantly Increasing $75

SMALL CAPS: 12-POINT, 9A, 50c.   10-POINT, 10A, 50c.
8-POINT, 12A, 45c.   6-POINT, 16A, 40c. EXTRA.

HHHHHHHHHHHHHHHH HHHHH HHHHH

1a 3A, $8.00 — 48-POINT EXTENDED OLD STYLE — L. C. $3.10; C. $4.90

FINISH
Perfect 6

5a 3A, $5.50 — 36-POINT EXTENDED OLD STYLE — L. C. $2.40; C. $3.10

DESIGN
Combine 18

5a 3A, $4.30 — 30-POINT EXTENDED OLD STYLE — L. C. $1.90; C. $2.40

INCREASE
Specimen 34

8a 4A, $3.50 — 24-POINT EXTENDED OLD STYLE — L. C. $1.80; C. $1.70

FILED AWAY
Bettering Values
Improved 25

12a 6A, $3.20   18-POINT EXTENDED OLD STYLE   L. C. $1.65; C. $1.55

MODERN SYSTEM
Fine Printing Material
Superior Types 12

18a 10A, $2.80   12-POINT EXTENDED OLD STYLE   L. C. $1.40; C. $1.40

COMPARE THE PRICES
Respond to Your Best Interest
Profitable Supplies 43

20a 14A, $2.50   10-POINT EXTENDED OLD STYLE   L. C. $1.20; C. $1.30

WRITE FOR ESTIMATE NOW
Stock of Printers' Goods is Complete
Every Department Filled 95

22a 14A, $2.25   8-POINT EXTENDED OLD STYLE   L. C. $1.10; C. $1.15

FURNISHED FINE PRODUCTIONS
STANDARD LINE TYPES
Extended Old Style Face Outranks its Class
Advance of the Procession $76

28a 15A, $2.00   6-POINT EXTENDED OLD STYLE   L. C. $1.00; C. $1.00

SPECIAL SMALL CAPS ARE UNNECESSARY
IMPORTANT FOR COMPOSITORS
Being Cast on Standard Line any Size Caps can be Used
as the Small Caps of the Size Next Larger 180

HHHHHH
HHHHHHHHHHHHHH

FULL-FACE SERIES

22a 14A, $2.80　　　　12-POINT FULL-FACE NO. 1　　　　L. C. $1.40; C. $1.40

IMPROVED PLAIN LETTERS
Old Stand-Bys and Favorites Bettered
We Cast All on Standard Line 250

28a 16A, $2.65　　　　11-POINT FULL-FACE NO. 1　　　　L. C. $1.30; C. $1.35

FULL-FACES BROUGHT INTO SYSTEM
Types for Titles, Side-Heads and Emphatic Words
Lining Accurately with Our Roman Faces 316

28a 16A, $2.50　　　　10-POINT FULL-FACE NO. 1　　　　L. C. $1.25; C. $1.25

DESIRABLE FOR TABULAR WORK
Figures Altered and Cast Systematic in Thickness
Are Fitted to Point Sets in All the Sizes $48

30a 16A, $2.40　　　　9-POINT FULL-FACE NO. 1　　　　L. C. $1.20; C. $1.20

BECOMES CONSTANTLY MORE POPULAR
Advantages of Our Standard Line Meriting Appreciation
Those Once Using Our Type Want No Other 279

30a 16A, $2.25　　　　8-POINT FULL-FACE NO. 1　　　　L. C. $1.10; C. $1.15

PRODUCES LABOR-SAVING MATERIAL
Modern Methods Introduced by the Inland Type Foundry
Outclassing the Sleeping Old-Fogy Concerns 158

38a 22A, $2.00　　　　6-POINT FULL-FACE NO. 1　　　　L. C. $1.00; C. $1.00

IMPORTANT FACT WHICH SHOULD NEVER BE FORGOTTEN
Single and Dotted Two-Point Brass Rule Lines with Every One of Our Faces
Easily Justified to Line by Use of Point-System Leads and Slugs $369

32a 18A, $2.00　　　　5½-POINT FULL-FACE NO. 1　　　　L. C. $1.00; C. $1.00

MAKING THINGS COMFORTABLE FOR THE COMPOSITOR
Vexatious Chopping-Up of Paper and Cardboard Now Relegated to the Past
Setting of Legal Blanks and Similar Forms is a Pleasant Work 815

26a 15A, $2.40　　　　9-POINT FULL-FACE NO. 2　　　　L. C. $1.20; C. $1.20

POINTER FOR THE PROFIT-SEEKERS
Investigate the Merits of Our Standard Line Type
Makes Earning of Money an Easier Task 290

HHHHHHHHHHHHHHHHHHH

26a 16A, $2.80   12-POINT CONDENSED NO. 1   L. C. $1.40; C. $1.40

OUR FACILITIES FOR MAKING
Type are of the Very Best and Cannot Fail to Please the Most Fastidious Person 374

32a 20A, $2.65   11-POINT CONDENSED NO. 1   L. C. $1.30; C. $1.35

OUR STANDARD JOB FACES ARE MORE
Uniform in Face and Justification Than any Other Make
Ornaments and Borders of all Kinds in Stock 45

32a 20A, $2.50   10-POINT CONDENSED NO. 1   L. C. $1.25; C. $1.25

ALL PRINTERS SHOULD EXAMINE
Our System and Then Compare it With That of Others
It Always Stands the Tests Satisfactorily 984

38a 22A, $2.40   9-POINT CONDENSED NO. 1   L. C. $1.20; C. $1.20

WE MANUFACTURE A COMPLETE LINE OF BRASS
Rule and our Jobbing Department is the Most Complete in the West
Superior Printing Material of all Kinds Quickly Supplied 62

36a 20A, $2.25   8-POINT CONDENSED NO. 1   L. C. $1.15; C. $1.10

ALL FACES ON EACH BODY, ROMANS, ITALICS,
Antiques, and all Job Faces Line With One Another at the Bottom
Our Figures Need no Special Justifying in Tabular Work

42a 26A, $2.20   7-POINT CONDENSED NO. 1   L. C. $1.10; C. $1.10

ALL OUR JOB TYPE CAN BE SET SOLID AND THE LIABILITY
Of the Descending Letters Breaking Off Is Entirely Overcome and all Printers Can See
That this in Itself is a Very Important Money-Saving Feature 295

44a 26A, $2.00   6-POINT CONDENSED NO. 1   L. C. $1.00; C. $1.00

THIS CONDENSED SERIES WILL LINE WITH OUR ROMAN
Without the use of Cardboard and can be Used in Sub-headings and in Tabular Work
Our Leaders Will Also Line With it as Well as With the Roman 790

HHHнппнHHHHHнппнHH

14a 7A, $3.50   24-POINT CONDENSED NO. 2   L. C. $1.80; C. $1.70

STANDARD LINE PRODUCT
Popular Among Printers 92

18a 10A, $3.30   20-POINT CONDENSED NO. 2   L. C. $1.65; C. $1.65

DEMANDS FOR MODERN SYSTEM
Now Constantly Increasing 56

20a 12A, $3.20   18-POINT CONDENSED NO. 2   L. C. $1.55; C. $1.65

INTRODUCING A UNIFORM METHOD
Product of Inland Type Foundry 35

22a 12A, $3.20   16-POINT CONDENSED NO. 2   L. C. $1.60; C. $1.60

SAVING IN BOTH TIME AND LABOR
Users of Standard Lining Faces 40

24a 15A, $3.00   14-POINT CONDENSED NO. 2   L. C. $1.50; C. $1.50

IMPROVEMENTS IN ORDINARY STYLES
Enhancing Their Usefulness Greatly 76

30a 18A, $2.80   12-POINT CONDENSED NO. 2   L. C. $1.40; C. $1.40

REMODELING COMMON FACES THROUGHOUT
Manifest Faults in Founding Remedied $34

34a 20A, $2.65   11-POINT CONDENSED NO. 2   L C. $1.30; C. $1.35

REPRODUCING POPULAR HANDSOME OLD-TIMERS
Supply Improvement in Many Essential Details 98

CONDENSED No. 2 SERIES

34a 20A, $2.50   10-POINT CONDENSED NO. 2   L. C. $1.25; C. $1.25

TAKING ADVANTAGE OF YEARS OF EXPERIENCE
Errors of Former Type Making are Corrected 50

38a 22A, $2.40   9-POINT CONDENSED NO. 2   L. C. $1.20; C. $1.20

CAREFUL CONSIDERATION GIVEN TO THE PRINTERS' WANTS
Best Labor-Saving Material the Result of Particular Study 49

38a 22A, $2.25   8-POINT CONDENSED NO. 2   L. C. $1.15; C. $1.10

STANDARD LINE AND UNIT SETS A DESIRABLE COMBINATION
Perfect Justification in Every Direction Easily Secured $82

38a 24A, $2.20   7-POINT CONDENSED NO. 2   L. C. $1.10; C. $1.10

REFORMS MADE THAT HAVE HITHERTO BEEN CONSIDERED IMPOSSIBLE
Obstacles Found to be Not so Insurmountable as Claimed by Founders 175

40a 25A, $2.00   6-POINT CONDENSED NO. 2   L. C. $1.00; C. $1.00

HINDERED BY NO PRECEDENTS OR TRADITIONS IN OUR TYPE FOUNDRY
Our House is Not Obliged to Follow the Old Methods and Systems 360

HHHHHHHH**H**HHHHHHHH

HALF-TITLE SERIES

25a 16A, $2.25   8-POINT HALF-TITLE   L. C. $1.15; C. $1.10

ALL OUR FACES LINE WITH ONE ANOTHER
Attaining Perfect Accuracy Without Use of Cardboard
Our Brass Rule Department is Very Complete 46

30a 18A, $2.00   6-POINT HALF-TITLE   L. C. $1.00; C. $1.00

IT WILL BE TO YOUR INTEREST TO BUY OUR TYPE
Standard Line Type Is the Only Material for Wide-Awake Printers
Elements of our Metal Combine Toughness and Hardness

28a 16A, $2.00   5½-POINT HALF-TITLE   L. C. $1.00; C. $1.00

LINING SYSTEM WILL SAVE LABOR IN JUSTIFICATION
We Buy and Sell Second-Hand Machinery, and can Often Offer you
Bargains in These Goods. Send for Illustrated Catalog 28

CONDENSED TITLE No. 2

10A   24-POINT CONDENSED TITLE NO. 2 $2.50

INCREASE USEFULNESS 16

14A   18-POINT CONDENSED TITLE NO. 2 $2.00

STANDARD LINE RECOMMENDED
PRINTERS INVESTIGATE 78

15A   16-POINT CONDENSED TITLE NO. 2 $1.80

MATERIAL MEETING THE DEMAND
THOROUGHLY REMODELED $840

15A   14-POINT CONDENSED TITLE NO. 2 $1.65

LATEST IDEAS IN TYPE FOUNDING
PROGRESSIVE INSTITUTION 95

16A   12-POINT CONDENSED TITLE NO. 2 $1.50

SUPERIOR IN WEARING QUALITIES TO
EVERY OTHER IN THE MARKET $48

20A   10-POINT CONDENSED TITLE NO. 2 $1.40

METAL COMPOSITION THE BEST EVER USED
OURS GUARANTEED MOST DURABLE 63

20A   9-POINT CONDENSED TITLE NO. 2 $1.35

INTERESTS PRINTERS THAT WORK FOR PROFITS
MOST PERFECT MATERIAL EVER OFFERED 25

8-POINT CONDENSED TITLE NO. 2   6-POINT CONDENSED TITLE NO. 2
24A, $1.25   25A, $1.00

MORE USEFUL ARE LEADERS   EXCELLENT PRODUCTIONS PLEASE
LINE WITH ALL FACES 70   SUPREMACY APPRECIATED 914

HHHHHHHHHHHHHHHHH

IONIC AND CLARENDON

16a 8A, $3.20   18-POINT IONIC   L. C. $1.65; C. $1.55

LEADING PRINTERS
Getting our Superior Type 28

24a 15A, $2.80   12-POINT IONIC   L. C. $1.40; C. $1.40

MODERN SYSTEMATIC PRODUCT
Setting the Pace for Our Competitors $30

34a 20A, $2.20   7-POINT IONIC   L. C. $1.10; C. $1.10

MONEY-SAVING TYPE MADE FOR SUPERIOR JOBBING
Great Possibilities in Having Systematic Material to Produce Fine Work
Point Bodies, Standard Line, Unit Sets and Most Useful Faces 279

34a 20A, $2.00   6-POINT IONIC   L. C. $1.00; C. $1.00

WE SUPPLY ALL FIGURES ON UNIFORM POINT SETS
They are also Multiples of and Justify with Our Regular Spaces and Quads
Advantages Inherent in the Unit-Set System of Casting Type 456

25A   7-POINT CLARENDON TITLE $1.20

ENDLESS VARIETY OF ORIGINAL BORDERS AND ORNAMENTS
MANY OF OUR BEAUTIFUL BORDERS ARE MADE FOR TWO COLORS
SPECIMEN SHEETS WILL BE FURNISHED ON APPLICATION $25

22A   6-POINT CLARENDON TITLE $1.00

ALL OUR TYPE CAN BE SET SOLID AND THE LIABILITY OF
THE DESCENDING LETTERS TO BREAK OFF IS ENTIRELY OVERCOME
OLD STYLE FIGURES MADE UNIFORM IN SIZE AND LINE 360

40a 24A, $2.20   7-POINT CLARENDON   L. C. $1.10; C. $1.10

PRINTERS ARE ADVISED TO BE CAREFUL IN BUYING TYPE
Bungling Imitations of Standard Line Type are Now Being Put on the Market
Not Made to Agree with our Accurate Dies and Measurements 487

38a 22A, $2.00   6-POINT CLARENDON   L. C. $1.00; C. $1.00

OUR STANDARD LINE SYSTEM MEETS ALL REQUIREMENTS
Every Possible Use of Type Has Been Carefully Considered and Years of Earnest
Study Were Devoted to Bringing It to the Acme of Perfection 19

4a 5A, $6.00　　　　42-POINT ANTIQUE NO. 1　　　　L. C. $2.40; C. $3.60

USEFUL
Products 5

5a 4A, $5.00　　　　36-POINT ANTIQUE NO. 1　　　　L. C. $1.90; C. $3.10

DESIGNS
Excellence 18

8a 4A, $4.30　　　　30-POINT ANTIQUE NO. 1　　　　L. C. $2.20; C. $2.10

FINER TYPE
Style Esthetic 12

12a 6A, $3.30　　　　20-POINT ANTIQUE NO. 1　　　　L. C. $1.65; C. $1.65

LINING SYSTEM
Problem Finds Solution
Standard Perfect 30

ANTIQUE SERIES

16a 9A, $3.20   18-POINT ANTIQUE NO. 1   L. C. $1.65; C. $1.55

PRINTERS REMEMBER
Standard Lining System Has
No Equal in the Market

26a 14A, $2.80   12-POINT ANTIQUE NO. 1   L. C. $1.40; C. $1.40

OUTFITS FOR NEWSPAPERS
Furnished on Short Notice and Estimates
Most Cheerfully Submitted 85

30a 16A, $2.50   10-POINT ANTIQUE NO. 1   L. C. $1.25; C. $1.25

SUPERIOR WORK AND LARGE PROFITS
Are a Natural Result of Using Standard Lining Type
Enterprising Printers all Recommend It 74

32a 16A, $2.25   8-POINT ANTIQUE NO. 1   L. C. $1.15; C. $1.10

BRASS RULE IN LABOR-SAVING FONTS
Circles, Ovals, Octagons and Many Other Special Shapes Are
Made to Order in the Shortest Possible Time 62

34a 20A, $2.20   7-POINT ANTIQUE NO. 1   L. C. $1.10; C. $1.10

COMPLETE OFFICES ALL ON STANDARD LINE SYSTEM
Number of Faces Cast on Uniform Line now so Large that the Printer
Need no Longer Buy the Old Money-Wasting Material 84

34a 20A, $2.00   6-POINT ANTIQUE NO. 1   L. C. $1.00; C. $1.00

REMEMBER, WE ARE AT ALL TIMES PREPARED TO
Prove Every Claim as Regards the Finish and Accuracy of our Product
In Type Making we Propose to Stay at the Very Top 36

HHHH HHHHHHHHHHHH hhHH

LATIN ANTIQUE SERIES

5a 3A $7.25   18-Point Latin Antique   L. C. $3.10; C. $1.15

REAPED
Cashiers 2

6a 4A, $5.00   36-Point Latin Antique   L. C. $2.20; C. $2.80

NEATEST
Ornaments 38

7a 5A, $1.30   30-Point Latin Antique   L. C. $2.00; C. $2.30

MONEY SURE
Return Profits 40

10a 6A, $3.50   24-Point Latin Antique   L. C. $1.75; C. $1.75

ARE JUSTIFIED
Point Set Figures With
Regular Spaces 75

EACH STANDARD LINE
Face Available for Date Lines
Use With 2-Point Rule

EXPERT COMPOSITORS DESIRE
Superior Material Wherewith to Execute
Commercial Work Most Rapidly

PROVED FINISH AND EXACTNESS
Of our Type Far Better than that of any Other
All Faces of One Body Will Line 53

OLD STYLE FIGURES PERFECTLY UNIFORM
They Agree With One Another in Height and Line, and
Cast to Point Sets, Justify with Spaces 210

FINEST DESIGNS IN ARTISTIC JOB FACES
Extensive and Splendid Exhibit of Original Ornaments
Foremost of High-Class Productions 79

STANDARD LINING TYPE   **WE ARE AGENTS FOR ALL**
Obviates Numerous Vexations   Kinds of Printing Machinery of
Keeps Temper Smooth 50   Latest Improved Styles 46

HHHh   hHHH
hHHHHHHHHHHHh

8a 5A. $5.00   36-POINT LATIN   L. C. $2.40; C. $2.60

TRUE SYSTEM
Accurate Faces 16

9a 6A. $4.30   30-POINT LATIN   L. C. $2.00; C. $2.30

IMPROVEMENTS
Original Type Designs
Appreciated 23

12a 8A. $3.50   24-POINT LATIN   L. C. $1.70; C. $1.80

POINT SYSTEM BOTH
Ways far Superior to the Old
Method of Justifying 45

18a 12A. $3.20   18-POINT LATIN   L. C. $1.55; C. $1.65

PERFECT JUSTIFICATION
Without Cardboards Thereby Saving
Large Profits to Printers 386

LATIN SERIES

28a 16A, $2.80   12-POINT LATIN   L. C. $1.45; C. $1.35

ESTIMATES FURNISHED TO PRINTERS
Brass Rule, Labor-Saving Fonts, Circles, Ovals
Other Special Shapes Made to Order $65

32a 20A, $2.50   10-POINT LATIN   L. C. $1.25; C. $1.25

HANDSOMEST ART ORNAMENTS AND BORDERS
An Endless Variety to Select From, Largest Assortment
Most Highly Finished Collection in the Market 240

32a 20A, $2.40   9-POINT LATIN   L. C. $1.20; C. $1.20

HIGHLY IMPROVED TYPE FACES DESIGNED
Material Supplied for Every Kind of Artistic Job Printing
Profits Assured on Work of Any Description 75

34a 20A, $2.25   8-POINT LATIN   L. C. $1.15; C. $1.10

GOOD WORK, LARGE PROFITS AND PLEASURE TOURS
Some of the Many Advantageous Results of Standard Line Type
Enterprising Printers Must Investigate this System 938

36a 24A, $2.20   7-POINT LATIN   L. C. $1.10; C. $1.10

LARGE ASSORTMENT OF FACES NOW CAST ON STANDARD LINE
Entirely Unnecessary for the Progressive Printer to Buy Type Not Systematic
Every Style We Produce Is Up-to-Date and Every Font Is Useful 210

36a 24A, $2.00   6-POINT LATIN   L. C. $1.00; C. $1.00

ALL SMALL CAP SORTS LIABLE TO BE CONFUSED WITH THE
Same Lower Case Characters Have an Extra Nick, Thereby Obviating Errors
Common in Distribution and Composition; Another Advantage 756

HHHHHHHHHHHHHHHHHH

Condensed Latin Series

8a 5A, $5.00 — 48-Point Condensed Latin — L. C. $2.40; C. $2.60

Beautiful ORNATE Art

10a 6A, $4.00 — 36-Point Condensed Latin — L. C. $2.00; C. $2.00

STANDARD Lining Type 95

18a 12A, $3.50 — 24-Point Condensed Latin — L. C. $1.70; C. $1.80

SUPERIOR AS TO DURABILITY
Accuracy and Finish of Our Products 25

30a 18A, $3.20 — 18-Point Condensed Latin — L. C. $1.60; C. $1.60

ONE OF THE MOST IMPORTANT INNOVATIONS
Is our Lining System and Printers Should Investigate 30

40a 26A, $2.80 — 12-Point Condensed Latin — L. C. $1.40; C. $1.40

OUR STANDARD JOBBING FACES ARE MORE UNIFORM
In Design, Justification and Finish than any Others in the Market 62

45a 30A, $2.50 — 10-Point Condensed Latin — L. C. $1.20; C. $1.30

REMEMBER THAT WE INVITE CAREFUL AND CRITICAL EXAMINATION
And are at all Times Prepared to Prove our Claims for Standard Line System $48

HHHHHHHHHHHHHHHHHHHH

NEAT CAPS
LIGHT FACE 6

6A 20-POINT BRUCE TITLE $2.25

EXCELLENCE
BEST MODELS 38

7A 16-POINT BRUCE TITLE $1.80 | 10A 12-POINT BRUCE TITLE $1.50

DESIGNS   TWO-LINES
NEWEST 5   OUR TITLES 4

12A 10-POINT BRUCE TITLE $1.40 | 14A 9-POINT BRUCE TITLE $1.35

ENGRAVERS   LODGE WORK
CHOICE PART 70 NICE PROGRAM 92

16A 8-POINT BRUCE TITLE NO. 2 $1.25 | 18A 6-POINT BRUCE TITLE NO. 1 $1.00

UNIFORM MATERIAL   NEAT SOCIETY PRINTING
MODERN ENTERPRISE 86   DELICACY AND SIMPLICITY 35

20A 6-POINT BRUCE TITLE NO. 2 $1.00 | 20A 6-POINT BRUCE TITLE NO. 3 $1.00

STANDARD LINE FACES   TWO-LINES CALLED TITLES
REMARKABLE PROGRESSION $74   OUR METHOD OF NAMING SERIES 612

EACH SIZE CAST ON SIX-POINT HAS A DIFFERENT NICK

8-POINT NO. 1 AND 6-POINT NO. 4 OF THIS SERIES IN PREPARATION

HHHHHHHH
HHHHHHHHHHH

SKINNER SERIES

Patent Pending

HAVING started in the most disas-
trous period that the type-founding
trade has ever experienced, the success
of the Inland Type Foundry is nothing
less than phenomenal. To-day, at the
expiration of a little over three years, its
business has many times exceeded that

INLAND TYPE FOUNDRY

Supplier of

PRINTERS' MATERIAL OF EVERY KIND

Manufacturer of

STANDARD LINE TYPE

Cast to Unit Sets

Art Ornaments and Borders

Brass Rule, Dashes, Circles, Etc.

SAINT LOUIS

which any other type foundry has ever
acquired in the most prosperous times.
This is largely attributable to Standard
Line, Unit-Set, and many other innovations which we have introduced in the
type, making our productions the most
desirable the practical printer can find.

INLAND ORNAMENTS, SERIES No. 25 --displayed above-- Per font, $2.00

6a 4A, $4.30   30-POINT SKINNER   L. C. $2.05; C. $2.25

HANDSOME
Neatest Faces 5

7a 4A, $3.50   24-POINT SKINNER   L. C. $1.75; C. $1.75

FINE LETTER
Program Enrich 14

10n 6A, $3.20   18-POINT SKINNER   L. C. $1.60; C. $1.60

ELEGANT MODELS
Finish Chaste Design 30

16a 10A, $3.00   14-POINT SKINNER   L. C. $1.50; C. $1.50

END OF THE CENTURY
Material Suited to the Time 26

12-POINT SKINNER		10-POINT SKINNER NO. 1	
20a 12A, $2.60	L. C. $1.40; C. $1.40	22a 14A, $2.50	L. C. $1.25; C. $1.25

DAINTY WORK   LEADING STYLE
Finest Obtained 15   Exquisite Display 48

8-POINT SKINNER		6-POINT SKINNER	
26a 15A, $2.25	L. C. $1.15; C. $1.10	30a 18A, $2.00	L. C. $1.00; C. $1.00

STANDARD LINING   SYSTEMATIC PRODUCTS
Modern Plan Introduced 36   Standard Line Type the Best $90

10-POINT SKINNER NO. 2 AND 9-POINT SKINNER IN PREPARATION.

HHHHHHHHHHH

4a 3A, $9.50   60-POINT WOODWARD   L. C. $3.70, C. $5.80

FORMS
Perfect 6

5a 4A, $7.25   48-POINT WOODWARD   L. C. $3.10; C. $4.15

BORDER
Specimen 8

7a 4A, $5.00   36-POINT WOODWARD   L. C. $2.40; C. $2.60

LINING TYPE
Money Saved 12

9a 5A, $4.30   30-POINT WOODWARD   L. C. $2.15; C. $2.15

BEST METHOD
Fixing Standards 20

9a 6A, $3.50   24-POINT WOODWARD   L. C. $1.60; C. $1.90

UNIFORM FIGURES
Casting Systematic 36

15n 9A, $3.20   18-POINT WOODWARD   L. C. $1.60; C. $1.60

ORDERS LINING TYPE
Exhibit of Common Sense 98

18a 12A, $3.00   14-POINT WOODWARD   L. C. $1.40; C. $1.60

STANDARD LINE PRODUCTION
Notice Constant Increase in Faces $15

22a 15A, $2.80   12-POINT WOODWARD   L. C. $1.35; C. $1.45

INVESTIGATION RECOMMENDED
Recognize Advantages of Standard Line 24

26a 16A, $2.50   10-POINT WOODWARD   L. C. $1.25; C. $1.25

SUPERIOR MATERIAL FOR PURCHASERS
Handsome Faces and Lining System Combined 16

28a 20A, $2.25   8-POINT WOODWARD   L. C. $1.05; C. $1.20

ALL PRINTERS' SUPPLIES ALWAYS IN STOCK
Thoroughly Equip Printing Establishments with Best Type 38

7-POINT WOODWARD	6-POINT WOODWARD
34a 20A, $2.20 L. C. $1.10; C. $1.10	34a 20A, $2.00 L. C. $1.00; C. $1.00
RIGHT QUALITIES IN TYPE	STANDARD LINE SYSTEM
Obtains Supply of Lining Faces 70	Type for the Up-to-Date Printer 12

HHHHHH HHHHHH HHHHHHHH

4a 3A, $9.50   60-POINT WOODWARD OUTLINE   L. C. $3.70; C. $5.80

FORMS
Perfect 6

5a 4A, $7.25   48-POINT WOODWARD OUTLINE   L. C. $3.10; C. $4.15

BORDER
Specimen 8

7a 4A, $5.00   36-POINT WOODWARD OUTLINE   L. C. $2.40; C. $2.60

LINING TYPE
Money Saved 12

9a 5A, $4.30   30-POINT WOODWARD OUTLINE   L. C. $2.15; C. $2.15

BEST METHOD
Fixing Standards 20

Woodward Outline

Patent Pending

9a 6A, $3.50 | 24-Point Woodward Outline | L. C. $1.60; C. $1.90

UNIFORM FIGURES
Casting Systematic 36
Regular in Widths

15a 9A, $3.20 | 18-Point Woodward Outline | L. C. $1.60; C. $1.60

ORDERS LINING TYPE
Exhibit of Common Sense 98
Keeping Up to Times

18a 12A, $3.00 | 14-Point Woodward Outline | L. C. $1.40; C. $1.60

STANDARD LINE PRODUCTION
Notice Constant Increase in Faces $15
Enterprising Young Foundries

22a 15A, $2.80 | 12-Point Woodward Outline | L. C. $1.35; C. $1.45

INVESTIGATION RECOMMENDED
Recognize Advantages of Standard Line 24
Introduce Labor-Saving System

Woodward and Woodward Outline are cast to the same widths, and one will register accurately over the other for use in two-color work.

HHH H H H HHHH
HHHH HHHH

5a 4A, $9.50   60-POINT CONDENSED WOODWARD   L. C. $4.10; C. $5.40

MIND RULE
Face Lined 10

8a 5A, $7.25   48-POINT CONDENSED WOODWARD   L. C. $3.55; C. $3.70

BUYER LIKES
Modern Prints 45

8a 6A, $5.00   36-POINT CONDENSED WOODWARD   L. C. $2.30; C. $2.70

SUPERIOR PRODUCT
Finest and Best Made 96

10a 6A, $4.30   30-POINT CONDENSED WOODWARD   L. C. $2.20; C. $2.10

USING STANDARD LINE
Saving Compositors' Labor 24

CONDENSED WOODWARD

Patented Aug. 4, 1896

12a 8A, $3.50   24-POINT CONDENSED WOODWARD   L. C. $1.70; C. $1.80

JUSTIFICATION MUCH SIMPLER
Furnish Every Face on Unit-Sets 78

16a 10A, $3.20   18-POINT CONDENSED WOODWARD   L. C. $1.60; C. $1.60

DOTTED BRASS RULE IN BLANK WORK
Easily Adjusted to Line with all Type Faces 26

22a 14A, $3.00   14-POINT CONDENSED WOODWARD   L. C. $1.50; C. $1.50

LEADERS LINE WITH ALL ROMAN AND JOB FACES
Abolish Buying Special Leaders for Many Display Fonts 40

28a 18A, $2.80   12-POINT CONDENSED WOODWARD   L. C. $1.40; C. $1.40

PROOF THAT THE IDEA OF UNIFORM LINING WAS FEASIBLE
Refusal of Older Founders to Cast Systematic Type Proving Stupidity 35
Desires of Progressive Printers Based on Experience

34a 22A, $2.50   10-POINT CONDENSED WOODWARD   L. C. $1.25; C. $1.25

ELEGANT CONDENSED DISPLAY FACE FOR NEWSPAPERS IS OFFERED
Handsome and Attractive Ornamentation of Headlines and Announcements $92
Suitable for Use in Every Specimen of First Class Typography

8-POINT CONDENSED WOODWARD 36a 25A, $2.25 L. C. $1.10; C. $1.15	6-POINT CONDENSED WOODWARD 48a 30A, $2.00 L. C. $1.00; C. $1.00
PERMITS PRINTERS TO SECURE PROFITS Labor-Saving Features of Standard Line Type 680 Notice the Quicker Results Obtained	**USEFUL MODERN FACE FOR SMALL BOX-HEADINGS** Condensed Letter for Crowding Matter into Little Space 475 Very Suitable for Time-Tables and the Like

4a 3A, $10.75   48-POINT EXTENDED WOODWARD   L. C. $4.15; C. $6.60

READ
Black 5

5a 3A, $6.40   36-POINT EXTENDED WOODWARD   L. C. $2.80; C. $3.60

SAMPLE
Charmer 8

5a 3A, $4.70   30-POINT EXTENDED WOODWARD   L. C. $2.10; C. $2.60

PERFECT
Good Form 6

7a 4A, $4.00   24-POINT EXTENDED WOODWARD   L. C. $2.00; C. $2.00

BROAD FACE
Extend Letter 12
Modern Plant

9n 5A, $3.20   18-POINT EXTENDED WOODWARD   L. C. $1.65; C. $1.55

BEST METHODS
Regularity Found 60
Improved Lining

14a 8A, $3.00   14-POINT EXTENDED WOODWARD   L. C. $1.50; C. $1.50

NEAT MODELS USED
Artistic Jobbing Letters 93
Useful and Desirable

16a 10A, $2.80   12-POINT EXTENDED WOODWARD   L. C. $1.40; C. $1.40

ECONOMICAL PRINTERS
Selecting Profitable Material 25
Buy Standard Line Type

18a 10A, $2.50   10-POINT EXTENDED WOODWARD   L. C. $1.25; C. $1.25

SETTING TYPE MADE EASIER
Annoying Hindrances Abolished $18
Introduced Improved Systems

8-POINT EXTENDED WOODWARD	6-POINT EXTENDED WOODWARD
22a 14A, $2.25 L. C. $1.10; C. $1.15	28a 16A, $2.00 L. C. $1.00; C. $1.00
MUCH SIMPLIFIED	QUICKER COMPOSING
Justification Perfected	With Standard Line System
Unit Set Types 34	Improved Methods 62

60-Point Extended Woodward in preparation.

HHHHHHHHHHHH
HHHHHHHHHHHHHH

GOTHIC No. 1 SERIES

5a 3A $6.00 — 42-Point Gothic No. 1 — L. C. $2.50; C. $3.50

HERALD
Advantage 6

6a 4A, $5.00 — 36-Point Gothic No. 1 — L. C. $2.20; C. $2.80

SCIENCE
Improved Type

8a 5A, $4.30 — 30-Point Gothic No. 1 — L. C. $2.00; C. $2.30

COMPOSING
Standard Lining 3

10a 6A, $3.50 — 24-Point Gothic No. 1 — L. C. $1.65; C. $1.85

PRINTER FINDS
Chance to Save Work 18

16a 9A, $3.20 — 18-Point Gothic No. 1 — L. C. $1.55; C. $1.65

LINING TYPE PAYS BIG
Dividends to Careful Buyers 50

INLAND TYPE FOUNDRY — 90 — ST. LOUIS, MO., U. S. A.

GOTHIC No. 1 SERIES

⌐⌐⌐

20a 12A, $3.00   14-POINT GOTHIC No. 1   L. C. $1.40; C. $1.60

MOST DURABLE BODIES
Made from Combination Metal 26

25h 14A, $2.80   12-POINT GOTHIC No. 1   L. C. $1.40; C. $1.40

ORIGINAL AND UNIFORM FACES
Characteristics of Standard Line Type 80

26a 14A, $2.50   10-POINT GOTHIC No. 1   L. C. $1.25; C. $1.25

GOTHICS AND TITLES IMPROVED
Standard Line Justifies Without Cardboard 3

30a 16A, $2.40   9-POINT GOTHIC No. 1   L. C. $1.20; C. $1.20

LABOR-SAVING TYPE BODIES ARE PERFECT
No Trouble to make Leaders Line with Every Face 913

34a 18A, $2.25   8-POINT GOTHIC No. 1   L. C. $1.15; C. $1.10

OUR TYPE METAL IS OF A NEW COMPOSITION
Guaranteed to be Superior in Finish and Durability to all Others
Highest Quality and the Best Workmanship 86

38a 20A, $2.20   7-POINT GOTHIC No. 1   L. C. $1.10; C. $1.10

EVERY COMPOSITOR APPRECIATES STANDARD LINE TYPE
Annoying Complications Peculiar to the Old Style Justification are Obviated
Vexatious Troubles Incident to Type-Setting Abolished 315

38a 20A, $2.00   6-POINT GOTHIC No. 1   L. C. $1.00; C. $1.00

PRINTERS SHOULD INVEST IN LABOR-SAVING SYSTEM
Modern Improved Material Designed for the Use of Enterprising Craftsmen
Desires of Progressive Typographers Have Been Met 26

HHHHHHHHHHH
HHHHHHH HHHHHH

TITLE GOTHIC No. 5

STANDARD LINE · ITF · INVENTED BY INLAND TYPE FOUNDRY

4A   36-Point Title Gothic No. 5 $3.50

GRINDERS

5A   30-Point Title Gothic No. 5 $3.00

BROKERAGE

7A   24-Point Title Gothic No. 5 $2.50

SAVE MATERIAL

9A   20-Point Title Gothic No. 5 $2.25

BUY STANDARD LINE

12A   16-Point Title Gothic No. 5 $1.80

NOTABLE INNOVATION
OBTAIN MODERN FACES 18

12A   12-Point Title Gothic No. 5 $1.50

JUSTIFICATION SIMPLIFIED
BEST SYSTEM OF WIDTHS 26

16A   10-Point Title Gothic No. 5 $1.40

STANDARD ORIGINAL GOTHICS
MOST ACCURATE IN THEIR LINING 5

TITLE GOTHIC No. 5

〜⌣〜

16A   9-POINT TITLE GOTHIC No. 5 $1.35

GOTHICS AND TITLES IMPROVED
MAKE MONEY WITH STANDARD LINE 10

18A   8-POINT TITLE GOTHIC No. 51 $1.25

OUR BODIES ARE ACCURATE AND PERFECT
UNIFORM FACES AND FIGURES ARE SECURED 56

22A   8-POINT TITLE GOTHIC No. 52 $1.25

GOOD WORK, LARGE PROFITS AND PLEASURE
ARE RESULTS OF OUR NEW STANDARD LINE SYSTEM 32

20A   6-POINT TITLE GOTHIC No. 51 $1.00

**OUR LEADERS LINE WITH ALL BODIES BY MEANS OF LEADS
BRASS RULES, BRACES, DASHES, CIRCLES AND OVALS IN STOCK
EVERY COMPOSITOR APPRECIATES OUR SYSTEM 43**

24A   6-POINT TITLE GOTHIC No. 52 $1.00

**ALL FACES ARE MADE TO LINE BY USE OF LEADS AND SLUGS
COMMERCIAL PRINTERS READILY SEE THE MANY ADVANTAGES SECURED
MONEY-MAKERS INVEST IN LABOR-SAVING MATERIAL 28**

28A   6-POINT TITLE GOTHIC No. 53 $1.00

WE ARE AGENTS FOR ALL MAKES OF PAPER CUTTERS AND PRINTING PRESSES
OUR STOCK OF CASES, CABINETS, STANDS AND ALL OTHER WOOD GOODS IS COMPLETE
USEFUL AND DURABLE TYPE ADMIRED BY PRUDENT PURCHASERS $915

30A   6-POINT TITLE GOTHIC No. 54 $1.00

PUT IN STANDARD LINE TYPE IN PLACE OF WHAT YOU ARE USING AND OBSERVE THE
ADVANTAGE YOU HAVE OVER COMPETITORS; NO CARDBOARD OR PAPER NEEDED WITH OUR TYPE
LABOR SAVED IS MONEY EARNED AND CAN BE DONE WITH OUR MATERIAL 478

EACH SIZE CAST ON SIX-POINT HAS A DIFFERENT NICK

6-POINT TITLE GOTHIC Nos. 52, 53 AND 54 WILL ALSO BE CAST TO ORDER ON 5-POINT
BODY, IN FONTS OF 10 POUNDS AND OVER, AT SECOND CLASS PRICES.

HHHHHHHH HHHHHHHHHH HHHHHHHHHHHHHHHHHHHH

GOTHIC No. 6 SERIES
Original

10a 5A, $3.50	24-Point Gothic No. 6	L. C. $1.80; C. $1.70

HANDSOME STYLES
Modern Patterns 25

16a 7A, $3.30	20-Point Gothic No. 6	L. C. $1.75; C. $1.55

COMMEND PERFECTION
Believed Admirable 63

25a 12A, $3.20	16-Point Gothic No. 6	L. C. $1.65; C. $1.55

DESIGNING ELEGANT FASHION
Neat Standard Line Products 14

26a 12A, $3.00	14-Point Gothic No. 6	L. C. $1.60; C. $1.40

SCIENTIFIC METHOD OF FOUNDING
Furnish Most Accurate Material 82

32a 16A, $2.80	12-Point Gothic No. 6	L. C. $1.50; C. $1.30

LEADERS MADE MORE USEFUL THAN EVER
Upon our System they Line with all Faces 36

34a 16A, $2.50	10-Point Gothic No. 6	L. C. $1.30; C. $1.20

NUMEROUS ADVANTAGES OF OUR STANDARD LINE
Recognized by all Intelligent Type Buyers $75

36a 20A, $2.25	8-Point Gothic No. 6	L. C. $1.15; C. $1.10

ENDORSING INNOVATIONS IN THE MANUFACTURE OF TYPE
Our Various Improvements Give Pleasure to all Printers 18

6-Point Gothic No. 6	5-Point Gothic No. 6
40a 22A, $2.00 L. C. $1.00; C. $1.00	42a 24A, $2.00 L. C. $1.00; C. $1.00

PROFITS SURE IN STANDARD LINE   ACCURATE INTERLINING IS READILY ACCOMPLISHED
Labor-Saving Ideas Proved   Cardboard or Paper Rendered Useless
Remove Doubts as to Expediency 35   These Troublesome Adjuncts to Justification $69

HHHHHHHHHHHHHHHHHHHHH

TITLE GOTHIC No. 7
Original

9A · 20-POINT TITLE GOTHIC NO. 7 · $2.25

IMPROVED DESIGNS

10A · 18-POINT TITLE GOTHIC NO. 7 · $2.00

WERE ALL REMODELED
HANDSOME TYPE 12

14A · 14-POINT TITLE GOTHIC NO. 7 · $1.65

MODERN SYSTEMS OF LINING
SUPERIOR PRODUCT 48

16A · 12-POINT TITLE GOTHIC NO. 7 · $1.50

EVERY SERIES ON STANDARD LINE
PRINTERS' MONEY-SAVER $50

20A · 10-POINT TITLE GOTHIC NO. 7 · $1.40

AVOIDS CUTTING CARDBOARD AND PAPER
ONLY METAL LEADS NEEDED 63

9-POINT TITLE GOTHIC NO. 7 · 20A, $1.35 | 8-POINT TITLE GOTHIC NO. 7 · 24A, $1.25

FOUND TO SAVE LABOR · SYSTEM IN FIGURE WIDTHS
BETTERED TYPE 45 · OBTAIN UNIFORMITY 98

6-POINT TITLE GOTHIC NO. 71 · 22A, $1.00 | 6-POINT TITLE GOTHIC NO. 72 · 28A, $1.00

CAST TO MULTIPLES OF SPACES · STANDARD LINE TYPE WITHOUT A RIVAL
FIGURES AND POINTS $60 · NO OTHER CAN SURPASS IT 35

6-POINT TITLE GOTHIC NO. 73 · 30A, $1.00 | 6-POINT TITLE GOTHIC NO. 74 · 35A, $1.00

USEFULNESS OF SYSTEMATIC TYPE APPRECIATED · COMBINATIONS OF VARIOUS SIZES AS CAPS AND SMALL CAPS
OUR INNOVATIONS IN LETTER FOUNDING 32 · EASILY MADE WITHOUT INTRICATE JUSTIFICATION $479

EACH SIZE CAST ON SIX-POINT HAS A DIFFERENT NICK

HHHHHHHHHHHHHHHHHHHHHH

5a 4A, $10.00   72-POINT CONDENSED GOTHIC No. 31   L. C. $1.10; C. $5.90

Gothic Face 9

THIS SIZE IS CAST ON STANDARD TITLE LINE.

5a 4A, $8.75   72-POINT CONDENSED GOTHIC No. 1   L. C. $3.50; C. $5.25

PERFECTION

Uniform Lines 6

6a 4A, $7.00   60-POINT CONDENSED GOTHIC No. 1   L. C. $3.40; C. $3.60

GOTHIC SERIES

Made Complete 5

Sa 4A, $6.40   54-POINT CONDENSED GOTHIC NO. 1   L. C. $3.25, C. $3.15

ELEGANT MODES
Perfected Letters 28

Sa 5A, $6.00   48-POINT CONDENSED GOTHIC NO. 1   L. C. $2.90; C. $3.10

IMPROVED GOTHIC
Largest Series Cast 10

9a 5A, $1.50   42-POINT CONDENSED GOTHIC NO. 1   L. C. $2.25; C. $2.25

ARTISTIC DESIGNING
Cut and Finish Superb 96

9a 6A, $4.00   36-POINT CONDENSED GOTHIC NO. 1   L. C. $1.90; C. $2.10

TYPE LINING SYSTEM
Appreciated by Printers 54

CONDENSED GOTHIC No. 1 SERIES
Original

12a 7A, $3.50   30-POINT CONDENSED GOTHIC NO. 1   L. C. $1.80; C. $1.70

COLLECTION OF FINE BORDERS
Display the Greatest Assortment 12

20a 10A, $3.20   24-POINT CONDENSED GOTHIC NO. 1   L. C. $1.65; C. $1.55

DEPARTMENTS ARE ALL EQUIPPED
With Latest and Most Improved Machinery 83

26a 16A, $3.00   18-POINT CONDENSED GOTHIC NO. 1   L. C. $1.50; C. $1.50

FULLY INVESTIGATE OUR STANDARD LINE TYPE
Grand System Long Desired by Progressive Printers 45

38a 20A, $2.80   14-POINT CONDENSED GOTHIC NO. 1   L. C. $1.40; C. $1.40

WE ARE AGENTS FOR VARIOUS PRINTING PRESSES
Our Stock of Printers' Supplies is Complete in Every Particular 160

12-POINT CONDENSED GOTHIC NO. 1	10-POINT CONDENSED GOTHIC NO. 1
44a 24A, $2.50 L. C. $1.25; C. $1.25	44a 25A, $2.50 L. C. $1.25; C. $1.25
ANOTHER IMPORTANT FEATURE	**STANDARD LINE GIVES YOU PROFITS**
Unit Sets Simplifying Composition 79	Labor and Money Saved with Best Type 82

8-POINT CONDENSED GOTHIC NO. 1	6-POINT CONDENSED GOTHIC NO. 1
40a 26A, $2.25 L. C. $1.10; C. $1.15	45a 28A, $2.00 L. C. $1.00; C. $1.00
UNIT SET SYSTEM HELPS COMPOSITORS	**SMALL SIZE CONDENSED GOTHIC DESIRABLE**
Manner of Casting Type in Respect to Width $36	Useful for Box-Headings and Railway Time-Tables 540

For Specimens of Larger Sizes See Preceding Pages

12a 7A, $3.50   24-POINT GOTHIC ITALIC NO. 1   L. C. $1.70; C. $1.50

SUPERIOR PRINTING
Elaborate Lining Effected 12

18a 10A, $3.20   18-POINT GOTHIC ITALIC NO. 1   L. C. $1.60; C. $1.60

ELEGANCE AND EXCELLENCE
Beautiful Designs in Decorative Art 5

28a 16A, $3.00   14-POINT GOTHIC ITALIC NO. 1   L. C. $1.50; C. $1.50

ALL FACES SHOULD LINE PERFECT
Lining System will Save Labor in Justification 13

34a 20A, $2.80   12-POINT GOTHIC ITALIC NO. 1   L. C. $1.40; C. $1.40

PRINTERS BUY STANDARD LINE TYPE ALWAYS
Attaining Perfect Accuracy Without the Need of Cardboard

36a 24A, $2.50   10-POINT GOTHIC ITALIC NO. 1   L. C. $1.20; C. $1.30

THE ANTI-TRUST TYPE FOUNDRY OF THE WEST
The Elements of our Metal Combine Toughness and Hardness 218

42a 25A, $2.25   8-POINT GOTHIC ITALIC NO. 1   L. C. $1.10; C. $1.15

OUR BRASS RULE DEPARTMENT IS THOROUGH AND COMPLETE
Borders and Ornamentation Devices in Most Elaborate and Endless Variety 83

46a 26A, $2.00   6-POINT GOTHIC ITALIC NO. 1   L. C. $1.00; C. $1.00

IN BUYING AN OUTFIT IT WILL BE TO YOUR INTEREST TO SEE US
Standard Line Type is the Only Material for Wide-Awake Printers to Handle and it Pays 45

HHHHHHHHH HHHHHHHHHHHHHHHH HHHHHHHHH

Condensed Title Gothic No. 3 Series

4A   72-Point Condensed Title Gothic No. 3 $6.00

MAIN KIND 9

5A   60-Point Condensed Title Gothic No. 3 $5.00

UNIFORMITY 4

5A   54-Point Condensed Title Gothic No. 3 $4.00

DARK PRINTS 26

6A   48-Point Condensed Title Gothic No. 3 $3.75

GREAT SYSTEM 30

6A   42-Point Condensed Title Gothic No. 3 $3.25

STANDARD PRIME 75

7A   36-POINT CONDENSED TITLE GOTHIC No. 3 $2.75

PROFITS CAPTURED 18

8A   30-POINT CONDENSED TITLE GOTHIC No. 3 $2.50

USES STANDARD LINE 95

24-POINT CONDENSED TITLE GOTHIC No. 3   20-POINT CONDENSED TITLE GOTHIC No. 3
12A, $2.25   15A, $2.00

SPEEDY WORK 2   LABOR IS EASIER 64

16-POINT CONDENSED TITLE GOTHIC No. 3   12-POINT CONDENSED TITLE GOTHIC No. 3
20A, $1.80   24A, $1.50

STANDARD LINING   POPULAR INVENTION
PRINTERS ADMIRING $60   GIVEN HEARTY WELCOME 73

10-POINT CONDENSED TITLE GOTHIC No. 3   9-POINT CONDENSED TITLE GOTHIC No. 3
25A, $1.40   28A, $1.35

GREATER STEP IN ADVANCE   IMITATION IS MOST FLATTERING
SECURING SYSTEMATIC TYPES 85   SEE RIVALS COPYING OUR METHODS 92

8-POINT CONDENSED TITLE GOTHIC No. 32   6-POINT CONDENSED TITLE GOTHIC No. 3
30A, $1.25   28A, $1.00

BEWARE OF POOR COUNTERFEITS   NOT ALL FOUNDERS KNOW OUR PLANS
FAULTY LINING METHODS ARE BEING USED 74   OUR STANDARD LINE SYSTEM MOST ACCURATE S125

HHHHHHHHHHHHHHHHHH
HHHHHHHHH HHHHHHH

GOTHIC SPECIMENS

STANDARD LINE — ITF — INVENTED BY INLAND TYPE FOUNDRY

5A   36-POINT CONDENSED TITLE GOTHIC No. 2 $2.50

PERFECTED SYSTEM
SUPERIORITY 50

30a 16A, $3.00   14-POINT CONDENSED GOTHIC No. 1   L. C. $1.55; C. $1.45

FIGURES OF ALL FACES CAST ON POINT-WIDTHS
Being Also Exact Multiples of Our Spaces 46

GOTHIC TITLE ITALIC NO. 2

10A   20-POINT TITLE GOTHIC ITALIC No. 2 $2.25

STYLISH ORNAMENTS
LARGEST ASSORTMENT 16

16-POINT TITLE GOTHIC ITALIC No. 2   12A, $1.50

CARDBOARD
NOT NEEDED 38

12-POINT TITLE GOTHIC ITALIC No. 2   16A, $1.50

JUSTIFICATION
MUCH SIMPLIFIED 40

10-POINT TITLE GOTHIC ITALIC No. 2   20A, $1.40

UNIFORMITY PLEASES
HIGHLY SATISFACTORY 25

8-POINT TITLE GOTHIC ITALIC No. 21   24A, $1.25

STANDARD LINE SYSTEM
CANNOT BE IMPROVED ON 198

6-POINT TITLE GOTHIC ITALIC No. 22   25A, $1.25

PINACLE OF PERFECTION
MOST CORRECT LINING DEVISED 74

6-POINT TITLE GOTHIC ITALIC No. 2   26A, $1.00

SHUN MISERABLE IMITATIONS
STANDARD LINE SYSTEM BEING COPIED 50

HHHHHH*H**H*HHHHHH   HHHHHH*H**H*HHHHHH

Title Gothic Slope

Improved Series

6A   24-Point Title Gothic Slope $2.50

LINED FACES

8A   20-Point Title Gothic Slope $2.25

STANDARD LINE

10A   16-Point Title Gothic Slope $1.80

BORDER ORNAMENTS
INVESTIGATION 82

12A   12-Point Title Gothic Slope $1.50

TITLE GOTHIC SLOPES ARE
BEST IN THE MARKET 15

15A   10-Point Title Gothic Slope $1.40

OUR ARTISTIC PRODUCTIONS WILL
DELIGHT THE PRINTERS 348

16A   8-Point Title Gothic Slope $1.25

**MONEY SAVED BY USING STANDARD LINING
TYPE FOR FINE COMMERCIAL WORK 70**

6-Point Title Gothic Slope No. 61   6-Point Title Gothic Slope No. 62
20A, $1.00   22A, $1.00

SUPERIOR ART NOVELTIES   **SAVE MONEY BY INVESTMENT**
FOR ARTISTIC WORK 18   **IN STANDARD LINE $920**

6-Point Title Gothic Slope No. 63   6-Point Title Gothic Slope No. 64
24A, $1.00   25A, $1.00

FORTY TO FIFTY PER CENT SAVED   **HAVE YOU TRIED STANDARD LINE, THE**
BY USING STANDARD LINE 47   **BEST LABOR-SAVING MEDIUM? 5**

EACH SIZE CAST ON SIX-POINT HAS A DIFFERENT NICK

HHHHHHHHHH**HHH**H**H**H **H**HH**H**HHHHHHH

5a 3A $12.50   60-POINT COSMOPOLITAN   L. C. $4.60; C. $7.90

Charmed Displays 8

5a 3A, $7.50   48-POINT COSMOPOLITAN   L. C. $2.90; C. $4.60

Artist Buy Neat Faces 3

6a 3A, $5.50   36-POINT COSMOPOLITAN   L. C. $2.30; C. $3.20

Standard Line Meeting Demands

5a 4A, $5.00   30-POINT COSMOPOLITAN   L. C. $2.25; C. $2.75

Original Novelties Every Letter Useful 12

Cosmopolitan Series

Patented April 14, 1896

10a 5A, $3.80   24-POINT COSMOPOLITAN   L. C. $1.80; C. $2.00

Your Compositors Pleased Superior Material for Rapidity Triumph of Grand System 40

12a 5A, $3.30   18-POINT COSMOPOLITAN   L. C. $1.50; C. $1.80

Roman and Display Letters Perfect Faces Line with One Another Improvement in Jobbing Types 18

22a 7A, $3.00   14-POINT COSMOPOLITAN   L. C. $1.60; C. $1.40

Having been most carefully thought out, and all possible ways of using type considered in advance, not a single objection can be made to the system 25

30a 8A, $3.00   12-POINT COSMOPOLITAN   L. C. $1.75; C. $1.25

Being cast on Standard Line, the Cosmopolitan, aside from its usefulness as a letter for circulars, is specially available for date-lines and blanks that require an Italic letter, on account of the fact that dotted rule can be readily lined with the face 36

10-POINT COSMOPOLITAN
34a 9A, $2.75   L. C. $1.60; C. $1.15

All faces on any one body cast on the Standard Line system line accurately with one another, and all faces on different bodies are easily combined as caps and small caps 4

8-POINT COSMOPOLITAN
36a 10A, $2.25   L. C. $1.25; C. $1.00

Four new sizes here shown, the 60-Point, 14-Point, 10-Point and 8-Point, are recent additions to this very popular series and will augment its usefulness greatly. The demand for the 14-Point body is increasing, and it will be included in every new series we make $78

mmmm   *mmmm*

mmmmmmmmMmmmmmm

6a 4A, $9.50   60-POINT CONDENSED STUDLEY   L. C. $4.50; C. $5.00

CONDENSED
New Models 2

5a 5A, $7.25   48-POINT CONDENSED STUDLEY   L. C. $3.55; C. $3.70

PRINTS NICER
Artistic Design 48

9a 6A, $5.00   36-POINT CONDENSED STUDLEY   L. C. $2.40; C. $2.60

EXTENSIVE SELECTION
Cast on Standard Line 92

10a 7A, $4.30   30-POINT CONDENSED STUDLEY   L. C. $2.15; C. $2.15

LINING TYPES WELCOME
System has Popular Esteem 35

12a 9A, $3.50   24-POINT CONDENSED STUDLEY   L. C. $1.70; C. $1.80

DEMANDED OUR IMPROVEMENTS
Thoughtful Printers Now Delighted 24

18a 12A, $3.20   18-POINT CONDENSED STUDLEY   L. C. $1.60; C. $1.60

MEDIUM WIDTH FACE OF THIS STYLE
Wider Series Named Studley in Preparation 60

24a 16A, $3.00   14-POINT CONDENSED STUDLEY   L. C. $1.50; C. $1.50

BEAUTIFUL CONDENSED LETTER JUST PRODUCED
First Display of this Novel Face Shows Complete Series 79

30a 20A, $2.80   12-POINT CONDENSED STUDLEY   L. C. $1.40; C. $1.40

COMPLETE STANDARD LINE PRINTING OFFICES POSSIBLE
Large and Elegant Collection of Systematically Cast Types Now Offered $82
Specimens of Most Useful and Desirable Faces Supplied

34a 22A, $2.50   10-POINT CONDENSED STUDLEY   L. C. $1.25; C. $1.25

FITTING CLOSE OF THE PRESENT CENTURY OF THE TYPOGRAPHIC ERA
Reaching the Pinnacle of Perfection in the Art of Manufacturing Printing Type 36
System Established Regulating Bodies, Faces, Lines and Widths

8-POINT CONDENSED STUDLEY	6-POINT CONDENSED STUDLEY
36a 25A, $2.25 L. C. $1.10; C. $1.15	48a 30A, $2.00 L. C. $1.00; C. $1.00
STANDARD LINING TYPE A MONEY-MAKER	LEADERS LINE WITH ALL OUR ROMAN AND JOB FACES
Greatly Expedites Work in the Composing Room 75	Forget Not this Reiterated Statement of a Very Important Fact 12
Worries of Justification Now Obviated	Usefulness of Leaders is thus Greatly Augmented

HHHHHHHHHHHHHHHHHHHHHHH
HHHHHHHHHHHHHHHHHHHHHHHH

1a 3A. $13.50   60-POINT INLAND   L. C. $5.30, C. $5.20

PLANS
Grace 6

4a 3A. $8.40   48-POINT INLAND   L. C. $3.20, C. $5.30

LEADER
Elegant 15

5a 3A. $5.50   36-POINT INLAND   L. C. $2.25; C. $3.25

BEAUTEOUS
Novel Style 38

7a 4A. $4.30   30-POINT INLAND   L. C. $2.10; C. $2.20

MADE USEFUL
Each Lettering 42

8a 4A, $3.50   24-Point Inland   L. C. $1.75; C. $1.75

UNIQUE STYLES
Ornate Bold Faces 40

10a 6A, $3.20   18-Point Inland   L. C. $1.60; C. $1.60

STANDARD LINE TYPE
Appreciated by Printers 95

15a 8A, $3.00   14-Point Inland   L. C. $1.50; C. $1.50

RECOGNIZED SUPERIORITY
Ease of Justification Remarkable 93

20a 10A, $2.80   12-Point Inland   L. C. $1.45; C. $1.35

BEST FIGURES FOR TABULAR WORK
Ours are All Cast to Multiples of Spaces $12

20a 12A, $2.50   10-Point Inland   L. C. $1.25; C. $1.25

LEADERS LINE WITH ALL JOB FACES
Provided Only Standard Line Type is Bought 64

25a 16A, $2.25   8-Point Inland   L. C. $1.10; C. $1.15

HANDSOME DESIGNS SHOWN FOR MODERN PRINTING
Supplied to Leaders of Fashion in Neat and Elegant Typography 78

32a 18A, $2.00   6-Point Inland   L. C. $1.00; C. $1.00

QUICK WORK POSSIBLE WITH OUR SYSTEMATIC MATERIAL
Labor-Saving and Ornamental Faces and Borders for Artistic Compositors $58

HHHHHHHHHHHHHHHHH

1a 3A, $13.50   60-POINT EDWARDS   L. C. $5.30; C. $8.20

SHAPE
Finest 5

1a 3A, $8.50   48-POINT EDWARDS   L. C. $3.20; C. $5.30

BOLDER
Engrave 4

5a 3A, $5.50   36-POINT EDWARDS   L. C. $2.25; C. $3.25

PRINT DARK
Made Black 16

7a 4A, $4.30   30-POINT EDWARDS   L. C. $2.10; C. $2.20

UNIFORM LINE
Large Demand 30

EDWARDS SERIES

Patent Pending

Sn 4A, $3.50   24-POINT EDWARDS   L. C. $1.75; C. $1.75

PERFECT LINING
Accurate Features 72

10a 6A, $3.20   18-POINT EDWARDS   L. C. $1.60; C. $1.60

SYSTEMATIC FIGURES
Uniform Width in Points 30

15a 8A, $3.00   14-POINT EDWARDS   L. C. $1.50; C. $1.50

EXCELLENCE GUARANTEED
Perfection in Standard Line Type 64

20a 10A, $2.80   12-POINT EDWARDS   L. C. $1.45; C. $1.35

MORE EXTENDED USE FOR LEADERS
Found Possible by Our Method of Casting 85

20a 12A, $2.50   10-POINT EDWARDS   L. C. $1.25; C. $1.25

PAGE-MAKERS IN LETTER DESIGNING
Supplying Models for Rival Concerns to Copy 90

28a 16A, $2.25   8-POINT EDWARDS   L. C. $1.10; C. $1.15

SOME FEATURES WHICH MUST NOT BE OVERLOOKED
Figures in Old Style Fonts are Made Uniform in Size and Line $32

32a 18A, $2.00   6-POINT EDWARDS   L. C. $1.00; C. $1.00

EXTRA NICKS DISTINGUISH OUR OLD STYLES FROM ROMANS
Small Caps Letters Similar to the Lower Case are Likewise Specially Nicked 48

HHHHHHHHHHHH
HHHHHHHHHHHH

5a 1A, $7.25   48-POINT KELMSCOTT   L. C. $2.90; C. $4.35

OLD STYLES
Popular Series 3

5a 4A, $5.00   36-POINT KELMSCOTT   L. C. $2.45; C. $2.55

ANCIENT FACES
Excellent in Design 2

10a 5A, $4.30   30-POINT KELMSCOTT   L. C. $2.10; C. $2.20

GRACEFUL INITIALS
Those used by Radtolt are
very effective models 14

12a 6A, $3.50   24-POINT KELMSCOTT   L. C. $1.65; C. $1.85

YE OLDE PERIOD LIKED
Copies of the ancient book faces
now meet with a demand 68

A large number of the Ornaments shown in our book are suitable for use with this face.

18a 9A, $3.20   18-POINT KELMSCOTT   L. C. $1.60; C. $1.60

UNIQUE MODEL IS HANDSOME
Earliest faces of type revived and given a
cordial reception by artistic printers ✸ 70

24a 12A, $3.00   14-POINT KELMSCOTT   L. C. $1.50; C. $1.50

COMBINING OLD AND NEW FANCIES
In type designs we aim to suit the tastes of each
school and individual, hoping to please all with
an extensive variety of useful productions ✸ 32

30a 15A, $2.80   12-POINT KELMSCOTT   L. C. $1.40; C. $1.40

READY TO SUPPLY THE CRAFT'S DEMAND
Owing to a large number of inquiries if we could furnish
Mr. Morris' famous style of letter cast on Standard Line,
we have concluded to do so, and herewith show specimens
of the type used on the books of the Kelmscott Press ✸ 24

30a 16A, $2.50   10-POINT KELMSCOTT   L. C. $1.25; C. $1.25

IMPROVEMENTS ADDED WITH STANDARD LINING
As the matter set in this style of type is generally set solid, printers
will be glad to know that the Kelmscott Series is decidedly unique
in its class, because of the fact that in none of its sizes is there to be
found a single letter that kerns, either at head, foot or sides ✸ $18

35a 18A, $2.25   8-POINT KELMSCOTT   L. C. $1.15; C. $1.10

SOME WORDS OF WARNING TO THE PRINTING FRATERNITY
Reports have come to us that the travelling salesmen of some of the foundries, in
the trust as well as out of it, who have not yet adopted the Standard Line system,
are claiming that the houses they represent are casting Standard Line type. ✸✸
We would caution printers to be on their guard against any spurious product 350

HHHHHHHHHHHHHHHHHHHHH

RADTOLT INITIALS

E have copied these Initials from books printed by Radtolt, who is said to have been the first to print Initials

HESE Initials are useful not only in connection with the Kelmscott, but may with equally fine effect be used to embellish pages set in other old style body letter

Per Single Letter, 50c.   60-POINT RADTOLT INITIALS   Font of 26 Letters, $10.00

Per Single Letter, 40c.   48-POINT RADTOLT INITIALS   Font of 26 Letters, $9.00

Per Single Letter, 35c.   36-POINT RADTOLT INITIALS   Font of 26 Letters, $8.00

VERY printer will be pleased to see that the ancient style of letter which was revived and made so noted by Mr. Morris and his Kelmscott Press, may now be procured cast on Standard Line.

Saint John Initials

✿ ✿ ✿

Per Single Letter, 50c.   72-POINT SAINT JOHN INITIALS   Font of 26 Letters, $11.00

Per Single Letter, 40c.   48-POINT SAINT JOHN INITIALS   Font of 26 Letters, $9.00

Saint John Series

The Original—Received the Patent

5a 3A, $9.50   60-POINT SAINT JOHN   L. C. $4.10; C. $5.40

PRINTED

Bold Quaint

7a 3A, $7.25   48-POINT SAINT JOHN   L. C. $3.70; C. $3.55

ORIGINATES

Unique Models

9a 4A, $5.00   36-POINT SAINT JOHN   L. C. $2.60; C. $2.40

MODERN STYLE

Beautiful Products 28

Enterprising System

INLAND TYPE FOUNDRY   116   ST. LOUIS, MO., U. S. A.

Saint John Series

Patented Oct. 29, 1895

12a 5A, $3.50   24-POINT SAINT JOHN   L. C. $2.00; C. $1.50

STANDARD LINING TYPE
Desired by Quick Compositors 17
Speediest in Results Achieved

16a 6A, $3.20   18-POINT SAINT JOHN   L. C. $1.80; C. $1.40

CUSTOMERS OF PRINTERS
Novel Types Have Pleased Very Highly 36
Secure Patronage of Tasteful People

14-POINT SAINT JOHN
25a 9A, $3.00   L. C. $1.75; C. $1.25

SERIES NOW POPULAR

Encouraged by a very large demand two new sizes are added of this fine letter 15

12-POINT SAINT JOHN
30a 10A, $2.80   L. C. $1.70; C. $1.10

BOYCOTT THE IMITATIONS

Since the Saint John Series was first brought out by us, several other foundries have advertized poorly executed copies of it $20

10-POINT SAINT JOHN
34a 12A, $2.50   L. C. $1.50; C. $1.00

FOR MONEY-SAVING PRINTERS

The gain in the time it takes to set up our Standard Line and Unit Set type amounts to more than enough in one year to pay for the total cost of the type; throw away your old type 63

8-POINT SAINT JOHN
40a 15A, $2.25   L. C. $1.35; C. $0.90

ECONOMICAL STANDARD LINE BENEFITS

You can discard all your out-of-date material, purchase a brand-new outfit of Standard Line type, do better printing, and yet, at the end of the year, after deducting the entire cost of the outfit, find that your composing-room shows a larger profit than it has ever shown before 48

All sizes of this series from 8-Point to 24-Point will be furnished to order in fonts of 25 pounds and over, at second-class prices.

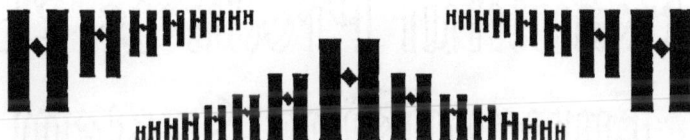

HHHHHHHH HHHHHHHH HHHHHHHH HHHHHHHH HHHHHHHH

Saint John Outline

Patent Pending

STANDARD LINE · ITF · INVENTED BY INLAND TYPE FOUNDRY

5a 3A, $9.50   60-POINT SAINT JOHN OUTLINE   L. C. $4.10; C. $5.40

PRINTED
Bold Quaint

7a 3A, $7.25   48-POINT SAINT JOHN OUTLINE   L. C. $3.70; C. $3.55

ORIGINATES
Unique Models

9a 4A, $5.00   36-POINT SAINT JOHN OUTLINE   L. C. $2.60; C. $2.40

MODERN STYLE
Beautiful Products 28
Enterprising System

Saint John Outline

Patent Pending

12a 5A, $3.50   24-POINT SAINT JOHN OUTLINE   L. C. $2.00; C. $1.50

STANDARD LINING TYPE
Desired by Quick Compositors 17
Speediest in Results Achieved

16a 6A, $3.20   18-POINT SAINT JOHN OUTLINE   L. C. $1.80; C. $1.40

CUSTOMERS OF PRINTERS
Novel Types Have Pleased Very Highly 36
Secure Patronage of Tasteful People

25a 9A, $3.00   14-POINT SAINT JOHN OUTLINE   L. C. $1.75; C. $1.25

MOST ARTISTIC OF OUTLINE STYLES
Encouraged by a Large Demand Two Sizes are Added
Appropriate Letter for Exquisite Job Work 15

30a 10A, $2.80   12-POINT SAINT JOHN OUTLINE   L. C. $1.70; C. $1.10

COMBINING ART, BEAUTY AND UTILITY
Enjoys Greatly the Success Achieved by the Inland Type Foundry
Modern Printer Gets the Benefit of Our Inventions 20

SAINT JOHN and SAINT JOHN OUTLINE are cast to the same widths, and one will register
accurately over the other for use in two-color work.

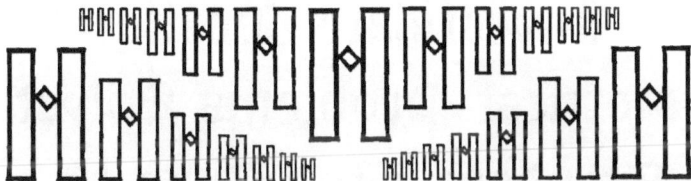

7n 3A, $7.25   48-POINT TUDOR BLACK   L. C. $3.60; C. $3.65

Lining Type
Desirable 8

10n 4A, $5.00   36-POINT TUDOR BLACK   L. C. $2.50; C. $2.50

Weekly Bargains
Line System 5

12n 4A, $4.30   30-POINT TUDOR BLACK   L. C. $2.40; C. $1.90

Easier Money Saving
Profits Assured 36

15n 5A, $3.50   24-POINT TUDOR BLACK   L. C. $2.00; C. $1.50

Investigate Merits
Knottiest Problems Solved
Standard Line Type 12

22a 8A, $3.20   18-Point Tudor Black   L. C. $1.75; C. $1.45

Perfect Justification Made Without any Cardboard or Paper Thereby Saving Large Profits

32a 10A, $2.80   12-Point Tudor Black   L. C. $1.60; C. $1.20

Estimates Furnished on Application
Brass Rule in Strips and Labor=Saving Fonts
Circles and Special Shapes to Order

36a 12A, $2.50   10-Point Tudor Black   L. C. $1.40; C. $1.10

Beautiful Art Ornaments and Original Borders
An Endless Variety to Select From for Newspapers
Magazines and Finest Artistic Job Work 14

44a 14A, $2.25   8-Point Tudor Black   L. C. $1.30; C. $0.95

Finest of Work, Largest Profits and Pleasure Journeys
Are Some of the Many Advantageous Results of Standard Lining
Enterprising Printers Investigate this System 23

50a 15A, $2.00   6-Point Tudor Black   L. C. $1.10; C. $0.90

All Small Cap Sorts Liable to be Confused With the Same Lower Case
Characters have an Extra Nick, Thereby Obviating Errors Common in Composition or
Distribution; Another Advantage to Proprietor and Compositor 6S

This is a very desirable face for German printing offices.
German accents are made for all sizes and furnished to order.

ฅฅฅฅฅฅฅฅฅฅฅฅ

MMMmmMmmmmmmmMM

~

50 pounds and over   36-POINT POSTER FRENCH OLD STYLE   50c. per pound

SUPREMACY
That the type of the Inland Type Foundry is finest all will admit 20

25 pounds and over   30-POINT POSTER FRENCH OLD STYLE   50c. per pound

SEE FIGURES
Note that all our Old Style figures are cast uniform in lining and agree in size on every body of each series 15

ALL POSTER FONTS CONTAIN A DUE PROPORTION OF SPACES AND QUADS.

〜

25 pounds and over   24-POINT POSTER FRENCH OLD STYLE   52c. per pound

SYSTEMATIC LINE
The value of our method of lining every type face has been acknowledged by all leading printers 36

25 pounds and over   20-POINT POSTER FRENCH OLD STYLE   52c. per pound

SIMPLIFYING WORK
By means of our Standard Line and Unit Set type printers can secure a great saving of labor in the composing departments 49

25 pounds and over   16-POINT POSTER FRENCH OLD STYLE   52c. per pound

PROGRESS OF TYPOGRAPHY
The Standard Line system is the greatest step forward that has been made in the present century of type founding, and it is very doubtful whether any additional improvement in the lining is possible 70

ALL POSTER FONTS CONTAIN A DUE PROPORTION OF SPACES AND QUADS.

25 pounds and over   24-POINT POSTER GOTHIC No. 6   52c. per pound

KNOWING MINDS
Printers no longer allow dealers to sell them type that is cast by the moss-backs on any old line 48

25 pounds and over   20-POINT POSTER GOTHIC No. 6   52c. per pound

POINTS AND LINES
It is true that there is no real value in any system of bodies unless it forms a basis for an ideal system of lining faces 15

25 pounds and over   16-POINT POSTER GOTHIC No. 6   52c. per pound

WHEREIN THE MERIT LIES
Even with point bodies the vexations incident to the use of cardboard and paper for justification cannot be done away with unless the faces are cast on a truly uniform lining system $26

ALL POSTER FONTS CONTAIN A DUE PROPORTION OF SPACES AND QUADS.

POSTER GOTHIC No. 6

25 pounds and over   14-POINT POSTER GOTHIC NO. 6   52c. per pound

MODERN STYLE OF GOTHIC
Not only are all our faces cast on Standard Line but they are all of the most approved modern cut, even the old stand-bys being remodelled and improved in appearance 30

25 pounds and over   12-POINT POSTER GOTHIC NO. 6   54c. per pound

BUY OUR STANDARD LINE LEADERS
They are made in four styles from 18-Point down to 5-Point; they will line not only with the Romans but also with all our job and display faces, a statement which none of the old out-of-date houses can make

25 pounds and over   10-POINT POSTER GOTHIC NO. 6   65c. per pound

MANIFOLD ADVANTAGES OF OUR SYSTEM
Printers recognize them at a glance, but were there no other reason the fact that perfect justification can always be had without cardboard or paper, and that no repeated trials are necessary to obtain good results, is enough to satisfy all 65

25 pounds and over   8-POINT POSTER GOTHIC NO. 6   80c. per pound

OUR INNOVATIONS IN THE MANUFACTURE OF TYPE
Such instant recognition and warm reception have been given to our new methods that all doubts as to their expediency and success have been removed. There is not a single printer to whom these improvements have been explained who has not heartily endorsed them $79

25 pounds and over   6-POINT POSTER GOTHIC NO. 6 $1.00 per pound

DO NOT CUT SPECIMEN BOOKS UNDER ANY CIRCUMSTANCES
This book is issued to our patrons to help them in making selections, and we would remind them that it is only necessary to give the Size, Name and Number, if any, of the Type, or Number of the Border, Cut or Rule desired, to insure the correct filling of any order. It is unnecessary to send lines, cuts or rules clipped from the book 315

ALL POSTER FONTS CONTAIN A DUE PROPORTION OF SPACES AND QUADS.

POSTER ANTIQUE No. 1

50 pounds and over   36-POINT POSTER ANTIQUE NO. 1   50c. per pound

PERFECT
Our Standard
Line meets all
lining needs 6

25 pounds and over   30-POINT POSTER ANTIQUE NO. 1   50c. per pound

IMPROVED
The favorite plain
faces are bettered
on this system 12

25 pounds and over   20-POINT POSTER ANTIQUE NO. 1   52c. per pound

MOST ACCURATE
No imitations can reach
Standard Line in giving
universal satisfaction 8

ALL POSTER FONTS CONTAIN A DUE PROPORTION OF SPACES AND QUADS.

〜

25 pounds and over   18-POINT POSTER ANTIQUE NO. 1   52c. per pound

SAVING OF LABOR
The type system that enables printers to economize in work is sure to give large profit 53

25 pounds and over   12-POINT POSTER ANTIQUE NO. 1   54c. per pound

MONEY MADE BY LINOTYPES
Some is made by help of the noted machine, but there is another which is not so limited in its capacity for general usefulness, and is recognized as our Standard Lineo'type $47

25 pounds and over   10-POINT POSTER ANTIQUE NO. 1   65c. per pound

SPECIALLY PROMINENT FEATURE
Do not overlook the fact that 2-Point single or dotted brass rules can be quickly justified in position to line with every one of our Standard Line faces, and makes the setting of legal or other blanks a pleasant work 38

25 pounds and over   8-POINT POSTER ANTIQUE NO. 1   80c. per pound

AUGMENTED USEFULNESS OF OUR LEADERS
Not only do the Leaders of each body line with every face we make on the same body, but they can very readily be used in connection with any Standard Line face cast on other bodies, the justification for accurate lining being simplicity itself 216

25 pounds and over   6-POINT POSTER ANTIQUE NO. 1 $1.00 per pound

SPURIOUS IMITATIONS OF STANDARD LINE TYPE
The perfection of our system was achieved only through many years of patient study and experiment, and the type we cast is made to agree with an elaborate, thorough and exact series of minute measurements and steel dies, none of which are in the hands of the pirating houses 90

ALL POSTER FONTS CONTAIN A DUE PROPORTION OF SPACES AND QUADS.

POSTER LATIN ANTIQUE

50 pounds and over   36-POINT POSTER LATIN ANTIQUE   50c. per pound

FASHIONS
No better type designs can be found than our house shows 9

30a SA   30-POINT POSTER LATIN ANTIQUE $12.50

DURABILITY
Of our type will in all cases give the Art Printer entire satisfaction. Give a thorough trial 3

ALL POSTER FONTS CONTAIN A DUE PROPORTION OF SPACES AND QUADS.

POSTER LATIN ANTIQUE

〰️

50a 10A   24-POINT POSTER LATIN ANTIQUE $12.50

LABOR-SAVING
Printers will save their money by purchasing Standard Line type to expedite their work 48

90a 20A   18-POINT POSTER LATIN ANTIQUE $12.50

OUR POSTER TYPE
All our type is cast Standard Line and all faces will line by the use of regular leads and slugs. This letter is specially available for poster work $15

160a 32A   12-POINT POSTER LATIN ANTIQUE $12.50

BOOK BINDERS INVESTIGATE
Our metal is a new composition, making type more durable; it prints better and gives finer results in electrotyping and stereotyping, and embossing, than any other. This interests book binders, as our faces will resist more pressure 612

ALL POSTER FONTS CONTAIN A DUE PROPORTION OF SPACES AND QUADS.

90a 20A   18-POINT POSTER IONIC $12.50

OUR LINING SYSTEM
Not only will the adoption of our system save enormously in labor and produce a better result, but as our material is more available for all classes of work a great saving of $25

150a 30A   12-POINT POSTER IONIC $12.50

IS STANDARD LINE A SUCCESS?
On January 1, 1893, we commenced doing business; in February, 1895, we issued our first small specimen book, and to-day the Inland Type Foundry is doing a business larger than any other foundry ever worked up in ten years. To what is this due if not to our superior system of lining faces? 36

25 pounds and over   6-POINT POSTER IONIC $1.00 per pound

USEFULNESS AND WORTH PROVED BY IMITATORS
The American type founders' trust, while not willing to acknowledge its indebtedness to a younger concern for the idea, has announced that it is making its new faces on a lining system. If it is a good thing for the new productions does it not follow that it is better still for the standard Full-Faces, Antiques and Gothics? Next, after having ridiculed the idea, a large Chicago foundry with almost unparalleled effrontery advertizes that its new faces are cast on "our" lining system, which it claims as original. But neither of the stolen systems is correct. Standard Line is the result of ten years of patient study and experiment, and when put into practice was found perfect. The imitators do not seem to know all the requirements 40

ALL POSTER FONTS CONTAIN A DUE PROPORTION OF SPACES AND QUADS.

50 pounds and over   36-POINT POSTER WOODWARD   50c. per pound

DELIGHTED
There is not one printer that has no kind word for Standard Line 8

25 pounds and over   30-POINT POSTER WOODWARD   50c. per pound

UNIFORMITY
Systematic material can always be relied upon to secure best results and give the most satisfaction 15

ALL POSTER FONTS CONTAIN A DUE PROPORTION OF SPACES AND QUADS.

25 pounds and over   24-POINT POSTER WOODWARD   52c. per pound

STYLISH SERIES
For posters the printer will not be able to get a more useful letter than this Woodward face 20

25 pounds and over   18-POINT POSTER WOODWARD   52c. per pound

PERFECT METHOD
Our Standard Line system is the only one that can possibly be devised to meet every one of the many requirements the true system has to satisfy 93

25 pounds and over   14-POINT POSTER WOODWARD   52c. per pound

SHUN FAULTY IMITATIONS
As there can be but one perfect lining system, we caution printers to beware of so-called "lining" faces put on the market by the old foundries; they will be found to be a vexatious delusion 46

ALL POSTER FONTS CONTAIN A DUE PROPORTION OF SPACES AND QUADS.

POSTER WOODWARD

Patented Aug. 4, 1896

25 pounds and over   12-POINT POSTER WOODWARD   54c. per pound

SECURE IDEAL IMPROVEMENTS

For many years intelligent printers have felt
and voiced the need of a systematic method
of lining all the various type faces, but until
the Inland Type Foundry came into the field
their calls for a reform in type-making have
fallen on the unwilling ears of old fogies 78

25 pounds and over   10-POINT POSTER WOODWARD   65c. per pound

FOLLOWING US AFTER A FASHION

Realizing at last the futility of ignoring further the
demand for a uniform lining system, the trust and
other founders have begun to experiment in this
direction, and are now offering abortive results to
the printer, under the claim that their type is cast
in accordance with their new "lining" system $10

25 pounds and over   8-POINT POSTER WOODWARD   80c. per pound

WHY NOT GET THE RIGHT KIND AT THE START?

Judging from their productions, these imitators are either too
ignorant to understand or too careless to meet all the requirements of the printer, and consequently their type has not the
advantages possessed by ours. If you buy their type you will
soon have to throw it away and get the genuine Standard Line
and Unit Set type. Our trade mark is on every package. 635

25 pounds and over   6-POINT POSTER WOODWARD $1.00 per pound

DO THE IMITATIONS COMBINE THESE QUALITIES?

The following are but a few of the features of our type: All Standard Line
faces on any one body line with one another; they line with all our faces on
other bodies with easy justification; the leaders of each body line with all
the job as well as Roman faces on that body; the leaders can be readily
justified to line with all faces on other than their own bodies; 2-Point
single or dotted brass rule is quickly lined with every face; every letter is
Unit Set. "Why buy the Second-Best when the Best costs no more?" 94

ALL POSTER FONTS CONTAIN A DUE PROPORTION OF SPACES AND QUADS.

50 pounds and over   36-POINT POSTER KELMSCOTT   50c. per pound

CHOICE KIND
Unique series cast on true lining system for first-class job letters 5

25 pounds and over   30-POINT POSTER KELMSCOTT   50c. per pound

MORRIS' STYLE
Peculiar type face revived by the noted poet's famous Kelmscott Press books 30

25 pounds and over   24-POINT POSTER KELMSCOTT   52c. per pound

MADE MORE USEFUL
By reason of its being placed on Standard Line this face is better than the other similar styles 16

ALL POSTER FONTS CONTAIN A DUE PROPORTION OF SPACES AND QUADS.

25 pounds and over   18-POINT POSTER KELMSCOTT   52c. per pound

SELECTION OF DESIGNS
The number and variety of faces that are cast on Standard Line enable every printer to buy the systematic type exclusively 24

25 pounds and over   14-POINT POSTER KELMSCOTT   52c. per pound

CHOICE COLLECTION OFFERED
In our specimen book may be found the cream of the old familiar faces as well as the best and most stylish creations in original faces, borders and ornaments suitable for every purpose ❦ 71

25 pounds and over   12-POINT POSTER KELMSCOTT   54c. per pound

ONE VERY PLEASING FEATURE
The changing from the old systems to the point system of type bodies caused confusion and trouble in all cases, but no such vexations follow the introduction of our Standard Line type into printing offices now using point bodies $38

25 pounds and over   10-POINT POSTER KELMSCOTT   65c. per pound

WORKS READILY WITH OTHER MATERIAL
While the most profit can be gained from Standard Line type by using it exclusively, the printer will find advantages in it even in the use of a single series, no trouble whatever being experienced in working it in jobs with the ordinary type of the older foundries 50

25 pounds and over   8-POINT POSTER KELMSCOTT   80c. per pound

MAKING ATTEMPTS TO MEET THE INEVITABLE
Finding that the general call for systematic lining type cannot be ignored, several of the old foundries, unwilling to come to a young concern for the perfect system, are now endeavoring to palm off ill-contrived "lining" systems of their own. Do we need to use strong language in putting you on your guard against them? $96

ALL POSTER FONTS CONTAIN A DUE PROPORTION OF SPACES AND QUADS.

Inland Type Foundry

Leads the Procession

Inventor of Standard Line Type

St. Louis, Mo., U. S. A.

STANDARD LINE TYPEWRITER

February 1, 1897

TO THE PROGRESSIVE PRINTER:

Dear Sir--We call your especial attention to our Type Metal, the result of years of careful research and experiment. This new composition we GUARANTEE to be harder, tougher and to wear longer than any other. Our equipment is the finest and most complete in the country, enabling us to turn out type, brass rule, etc., most accurately and promptly. All our tools are of our own design and built in our own shops, our casting machines being of an entirely new model capable of producing harder and more accurate type than those of other concerns. Remember ours is the only type foundry in the South and west of the Mississippi outside of the Trust.

Very truly yours,

INLAND TYPE FOUNDRY

St. Louis, Mo., U. S. A.

136

12-Point Typewriter
$60A,20A, including Spaces, $6.40

L. C. $4.10; C. $2.30

INLAND TYPE FOUNDRY

6-POINT GERMAN No. 1

THE STANDARD LINING SYSTEM

Eine höchst wichtige Neuerung in der Schriftgießerei ist das von uns erfundene und zuerst in der Inland Type Foundry unter dem obigen Namen eingeführte System gleicher Größe gegenseitig unter einander. Eine flüchtige Uebersicht der verschiedenen Schriftproben der letzten Jahre zeigt, daß von allen Seiten die Nothwendigkeit für die Uebereinstimmung der verschiedenen Schriften in Linie erkannt und viele dahinzielende Versuche gemacht worden sind, die aber leider sämmtlich als nicht zureichend sich herausgestellt haben. Alle unsere Schriften sind auf Standard Line gegossen und deßhalb halten alle Schriften auf selben Regel, sowohl Fraktur wie Antiqua, ebenso Cursiv und alle fetten, halbfetten und Zierschriften mit einander genaue Linie. Die Vorzüge dieses Systems sind so mannigfaltig, daß es unmöglich ist sie alle anzuführen, und wollen wir nur die folgenden erwähnen: Man kann alle Schriften der verschiedenen Gattungen, ebenso die verschiedenen Ziffern ohne alle Mühe verwenden und besonders in deutschen Druckereien können deßhalb für deutschen und englischen Satz dieselben Ziffern gebraucht werden. Letzteres ist besonders bei Tabellen für Eisenbahnen, wo verschiedene Ziffern in Anwendung kommen, von großem Vortheil. Selbst Schriften von verschiedenen Gattungen können auf unsere genaue Linie gebracht werden, und da unsere Schriften genau auf Punkt-System gegossen sind, können dieselben bei nicht comprespen Satz leicht als Durchschuß in Verwendung kommen. Auf diese Weise kommen z. B. Brevalia als Capitalchen der nächsten Größen dienen. Nicht allein stimmen alle Schriften in Linie mit STANDARD LEADERS, sondern sind ebenfalls in genauer Uebereinstimmung mit den einfachen oder punktirten Stücklinien in dem Gebrauch mit 2-Punkt oder 1-Punkt leads, und Normal-Quadraten. Trotz der praktischen Anwendung von leaders wird es hinwieder nothwendig, punktirte Stücklinien zu gebrauchen, und bei diese Neuerung, welche den Setzer ermöglicht, genaue Linie zu halten ohne das Flicken mit Kartenpapier, wird

abcdefghijklmnopqrstuvwxyz

INLAND TYPE FOUNDRY

8-POINT GERMAN No. 1

THE STANDARD LINING SYSTEM

Eine höchst wichtige Neuerung in der Schriftgießerei ist das von uns erfundene und zuerst in der Inland Type Foundry unter dem obigen Namen eingeführte System gleicher Linie in allen verschiedenen Schriften einer gleicher Größe gegenseitig unter einander. Eine flüchtige Uebersicht der verschiedenen Schriftproben letzter Jahre zeigt, daß von allen Seiten die Nothwendigkeit für Uebereinstimmung der verschiedenen Schriften in Linie erkannt, und viele dahin zielende Versuche gemacht worden sind, die aber leider alle als nicht zureichend sich herausgestellt haben. Alle unsere Schriften sind auf Standard Line gegossen und deßhalb halten alle Schriften auf denselben Regel, sowohl in Fraktur wie in Antiqua, ebenso Cursiv und alle fetten, halbfetten und Zierschriften mit einander genaue Linie. Die Vorzüge dieses Systems sind so mannigfaltig, daß es unmöglich ist sie alle anzuführen, und wollen wir nur die folgenden erwähnen: Man kann alle Schriften der verschiedenen Gattungen, ebenso verschiedene Ziffern ohne alle Mühe verwenden und deßhalb besonders in den deutschen Druckereien können für deutschen und englischen Satz dieselben Ziffern gebraucht werden. Letzteres ist besonders bei Tabellen für Eisenbahnen, wo verschiedene Ziffern in Anwendung kommen, von großem Vortheil. Selbst Schriften von verschiedenen

abcdefghijklmnopqrstuvwxyz

ST. LOUIS, MO., U.S.A.

THE STANDARD LINING SYSTEM

Eine in der Schriftgießerei höchst wichtige Neuerung ist das von uns erfundene und zuerst in der Inland Type Foundry unter den obigen Namen eingeführte System derselben Linie aller verschiedenen Schriften einer Größe gegenseitig unter einander. Eine flüchtige Uebersicht der verschiedenen Schriftproben letzter Jahre zeigt, daß von allen Seiten die Nothwendigkeit für eine Uebereinstimmung der verschiedenen Schriften in Linie erkannt, und viele dahin zielende Versuche gemacht worden sind, die aber leider sämmtlich als nicht zureichend sich erwiesen haben. Alle unsere Schriften sind auf Standard Line gegossen und deshalb halten alle Schriften auf jedem Regel sowohl Fraktur wie Antiqua, ebenso Cursiv und alle fetten, halbfetten sowie auch Zierschriften mit einander genaue Linie. Die Vorzüge des neuen Systems sind so mannigfaltig, daß es unmöglich ist, sie alle anzuführen, und wollen wir nur die folgenden erwähnen: Man kann alle Schriften der verschiedenen Gattungen, ebenso verschiedene Ziffern ohne alle Mühe verwenden und deshalb

abcdefghijklmnopqrstuvwxyz

STANDARD LINING SYSTEM

Eine besonders wichtige Neuerung in der Schriftgießerei ist das von uns erfundene und zuerst in der Inland Type Foundry unter dem obigen Namen eingeführte System gleicher Linie aller verschiedenen Schriften derselben Größe gegenseitig unter einander. Eine flüchtige Uebersicht der verschiedenen Schriftproben der letzten Jahre zeigt, daß von allen Seiten die Nothwendigkeit für die Uebereinstimmung der verschiedenen Schriften in Linie erkannt, und viele dahin zielende Versuche gemacht worden sind, die aber leider alle als nicht zureichend sich erwiesen haben. Alle unsere Schriften sind auf Standard Line gegossen und daher halten alle Schriften auf jedem Regel, sowohl Fraktur wie Antiqua, Cursiv und alle fetten, halbfetten und verschiedene Zierschriften mit einander genaue Linie. Die großen Vorzüge dieses Systems sind so mannigfaltig, daß es fast unmöglich ist, sie alle anzuführen, und wollen wir nur die folgenden

abcdefghijklmnopqrstuvwxyz

STANDARD LINING SYSTEM

Eine besonders wichtige Neuerung in der Schriftgießerei ist das von uns erfundene und zuerst in der Inland Type Foundry unter dem obigen Namen eingeführte System gleicher Linie aller verschiedenen Schriften derselben Größe gegenseitig unter einander. Eine flüchtige Ueberficht der verschiedenen Schriftproben der letzten Jahre zeigt, daß von allen Seiten die Nothwendigkeit für die Uebereinstimmung der verschiedenen Schriften in Linie erkannt, und viele dahin zielende Versuche gemacht worden sind, die aber leider alle als nicht zureichend sich erwiesen haben. Alle unsere Schriften sind daher halten alle Schriften auf selben Kegel, sowohl Fraktur wie Antiqua, Cursiv und alle fetten, halbfetten und verschiedene Zierzüge dieses Systems sind Linie. Die großen Vorzüge dieses Systems sind

abcdefghijklmnopqrstuvwxyz

STANDARD LINING TYPE

Eine höchst wichtige Neuerung in der Schriftgießerei ist das von uns erfundene und zuerst in der Inland Type Foundry unter obigen Namen eingeführte System gleicher Linie der verschiedenen Schriften einer Größe gegenseitig unter einander. Die flüchtige Ueberficht der verschiedenen Schriftproben letzter Jahre zeigt, daß von allen Seiten die Nothwendigkeit für die Uebereinstimmung der verschiedenen Schriften in Linie erkannt, und viele dahin zielende Versuche gemacht worden sind, die aber leider alle als unzureichend sich herausgestellt haben. Alle unsere Schriften sind genau auf Standard Line gegossen und deßhalb halten sämmtliche

abcdefghijklmnopqrstuvwxyz

German No. 1, Card Fonts

~

24a 8A. $2.80 12-POINT GERMAN NO. 1 L. C. $1.60; C. $1.20

Syſtematiſche Buch= und Accidenz=Schriften ausgeſtellt
Große Auswahl der allerbeſten Sorten $25

30a 10A. $2.50 10-POINT GERMAN NO. 1 L. C. $1.50; C. $1.00

Findet moderne Verbeſſerungen in dem Gebiete der Typographie
Erleichterte Arbeit für heutige Nachfolger Gutenbergs 36

30a 10A, $2.40 9-POINT GERMAN NO. 1 L. C. $1.45; C. $0.95

Eleganteſte und lesbarſte Fraktur=Schrift für Bücher und Zeitungen
Ein Triumph in der Kunſt des Stempelſchneiders 95

36a 12A, $2.25 8-POINT GERMAN NO. 1 L. C. $1.40; C. $0.85

Zufriedenheit aller Drucker mit dem von uns eingeführten Linien=Syſtem
Allgemeines Lob der Schriftſetzer zeigt deſſen großen Werth 48

42a 14A, $2.00 6-POINT GERMAN NO. 1 L. C. $1.20; C. $0.80

Alles Flicken mit Papier=Spähne im Juſtiren zweier Kegel mit einander iſt jetzt unnöthig
Genaue Stimmung der Linien erzielt ohne bekannte Schwierigkeiten 70

⋙ ⋘

German Full=Face Specimens

~

30a 10A, $2.50 10-POINT GERMAN FULL-FACE NO. 2 L. C. $1.45; C. $1.05

Höchſt exacte Uebereinſtimmung der Linien unſerer Schriften
Keine Ausnahmen in dem geſammten Sortiment 38

30a 10A, $2.40 9-POINT GERMAN FULL-FACE NO. 1 L. C. $1.40; C. $1.00

Nachfrage der modernen Drucker für ſyſtematiſches Material
Ekonomie mit gutem Produkt iſt ihr Wunſch 95

38a 14A, $2.20 7-POINT GERMAN FULL-FACE NO. 1 L. C. $1.25; C. $0.95

Die Vorzüge unſeres Syſtems ſind ſo mannigfaltig, daß es unmöglich iſt ſie alle
anzuführen; eine ſtrenge Unterſuchung iſt dem Buchdrucker empfohlen 26

38a 14A, $2.00 6-POINT GERMAN FULL-FACE NO. 1 L. C. $1.15; C. $0.85

Indem ſie in gleicher Linie ſtehen, können in deutſchen Druckereien dieſelben Ziffern
für deutſchen und engliſchen Satz ſehr vortheilhaft gebraucht werden 40

INLAND TYPE FOUNDRY 140 ST. LOUIS, MO., U. S. A.

Schwabacher Series

~~~

30a 10A, $2.80   12-POINT SCHWABACHER   L. C. $1.70; C. $1.10

Höchst wichtige Neuerung ist dem Buchdrucker vorgelegt
Das systematische Gießen aller Typensorten 28

40a 12A, $2.65   11-POINT SCHWABACHER   L. C. $1.60; C. $1.05

Alle unsere Schriften auf selben Kegel, obgleich Antiqua, Kursiv
oder Fraktur, stehen miteinander in genauer Linie $93

40a 12A, $2.50   10-POINT SCHWABACHER   L. C. $1.50; C. $1.00

Ziffern und Interpunktion stimmen in der Breite mit den Spatien
Sind gleich dem Kegel auf Punkt=System gegossen 46

42a 14A, $2.40   9-POINT SCHWABACHER   L. C. $1.40; C. $1.00

Eine Uebersicht der verschiedenen Probe=Blätter der letzten Jahre zeigt, daß
die Nothwendigkeit der Linien=Uebereinstimmung erkannt war 510

42a 14A, $2.25   8-POINT SCHWABACHER   L. C. $1.35; C. $0.90

Doch haben sich die vielen dahinzielende Versuche bis jetzt als nicht zureichend
hergestellt; die Lösung des Problems ist uns nun gelungen 372

→→→❯❮←←

Condensed German No. 1

~~~

42a 14A, $2.40   9-POINT CONDENSED GERMAN NO. 1   L. C. $1.40; C. $1.00

Schriften von verschiedenem Kegel und von jeder Gattung können mittels Ein=
und Zwei=Punkt Durchschluß leicht in genaue Linie gebracht werden 36

42a 14A, $2.25   8-POINT CONDENSED GERMAN NO. 1   L. C. $1.30; C. $0.95

Einfache und punktirte Messing=Linien auf Zwei=Punkt Kegel stimmen genau in
Linie mit unseren Schriften im Gebrauch von Normal=Durchschluß 45

50a 15A, $2.20   7-POINT CONDENSED GERMAN NO. 1   L. C. $1.30; C. $0.90

Trotz der praktischen Anwendung von punktirten Quadraten wird es öfters nothwendig, punktirte
Messing=Linien zu gebrauchen, deren Justirung unter unserem System erleichtert ist $70

50a 15A, $2.00   6-POINT CONDENSED GERMAN NO. 1   L. C. $1.20; C. $0.80

Die Aufmerksamkeit der Buchdrucker ist verlangt für die Proben unserer Fraktur=Schrift, welche
Gattung wir für die feinste und lesbarste halten, die noch je geschnitten worden ist 98

5a 3A. $9.50   60-POINT PREETORIUS   L. C. $4.10; C. $5.40

Journalist
Annoncen 3

7a 3A. $7.25   48-POINT PREETORIUS   L. C. $3.70; C. $3.55

Druckereien
Neue Schrift 18

9a 1A, $5.00   36-POINT PREETORIUS   L. C. $2.60; C. $2.40

Kräftige Lettern
Tägliche Zeitungen 36
Schwartz-Künstler

Preetorius Series

Patented Oct. 29, 1895

12a 5A, $3.50   24-POINT PREETORIUS   L. C. $2.00; C. $1.50

Moderne Weisen eingeführt
Systematische Linie für Typen 10
Elegante verbesserte Schriften

16a 6A, $3.20   18-POINT PREETORIUS   L. C. $1.80; C. $1.40

Eine höchst wichtige Neuerung
Nothwendigkeit der Uebereinstimmung 24
Guss der Schriften auf gleicher Linie

25a 9A, $3.00   14-POINT PREETORIUS   L. C. $1.75; C. $1.25

Mannigfaltige Vorzüge des neuen Systems
Gelobt von jedem der mit demselben bekannt wird 96
Alles Flicken mit Kartenpapier ist jetzt abgethan

30a 10A, $2.80   12-POINT PREETORIUS   L. C $1.70; C. $1.10

Schriften von verschiedem Kegel können mittels
Ein- und Zwei-Punkt Durchschuss in Linie gebracht werden 175
Unsere Spatien sind genau auf Punkt-Breiten gegossen

10-POINT PREETORIUS   8-POINT PREETORIUS
34a 12A, $2.50   L. C. $1.50; C. $1.00   40a 15A, $2.25   L. C. $1.35; C. $0.90

In deutschen Druckereien kann man für Fraktur und Antiqua Satz dieselben Ziffern gebrauchen, da unsere sämmtlichen Schriften selben Kegels auf gleicher Linie gegossen sind 48

Nicht allein stimmen alle unsere Schriften in Linie mit Standard Line Leaders, sondern sind ebenfalls in genauer Uebereinstimmung mit den einfachen und punktirten Messinglinien in dem Gebrauch mit 2-Punkt oder 1-Punkt Durchschuss und Normal-Quadraten. $3579

mmmMMmmm
MMmMmm mmmMmM

CONDENSED GOTHIC NO. 4
Manufactured by the Standard Type Foundry

10a 6A, $4.75   36-Point Condensed Gothic No. 4   L. C. $2.25; C. $2.50

PUREST Sentiment 19

12a 8A, $3.35   30-Point Condensed Gothic No. 4   L. C. $1.60; C. $1.75

TIN TOYS for Grown Babies 7

16a 10A, $3.05   20-Point Condensed Gothic No. 4   L. C. $1.50; C. $1.55

SURE FAITH CURE for Unbelievers 1

20a 10A, $2.60   18-Point Condensed Gothic No. 4   L. C. $1.40; C. $1.20

AMERICAN PATRIOTS Great Trouble 4

30a 16A, $3.00   14-Point Condensed Gothic No. 4   L. C. $1.55; C. $1.45

GOLDEN INDUCEMENTS Offered to Mechanics 96

12-Point Condensed Gothic No. 4   10-Point Condensed Gothic No. 4
20a 15A, $1.85   L. C. $1.00; C. $0.85   40a 25A, $2.60   L. C. $1.25; C. $1.35

HANDSOME Proprietors 13   MANAGE MUSIC Equal as Well 9

8-Point Condensed Gothic No. 4   6-Point Condensed Gothic No. 4
50a 30A, $2.30   L. C $1.15; C. $1.15   50a 30A, $1.95   L. C. $0.95; C. $1.00

YELLOW FEVER IS Coming this Year 11   COMPETITION SEASONED with Experiences 645

CONDENSED LINING GOTHIC
Manufactured by the Pacific States Type Foundry

32A   12-Point Condensed Lining Gothic No. 5 $1.75

ENERGETICALLY SPIN WITH BLOOMER GIRL DIPLOMATICALLY 1234

10-Point Condensed Lining Gothic No. 4   8-Point Condensed Lining Gothic No. 3
40A, $1.50   50A, $1.45

EMPHATICALLY MOUNT MY WHEEL 32   THEORETICAL PROBLEM FOR THINKERS 198

6-Point Condensed Lining Gothic No. 2   6-Point Condensed Lining Gothic No. 1
60A, $1.25   06A, $1.25

CLOUDS MAY FROWN AND THE WINDS MAY TWIRL 4563   YEARS MAY COME AND THE YEARS MAY GO AND THE PLANETS 378

INLAND TYPE FOUNDRY   144   ST. LOUIS, MO., U. S. A.

CUBS, SCROLLS, BIKES

Manufactured by the Pacific States Type Foundry

PACIFIC CUBS — Per font, $2.00; Single character, 25c.

PACIFIC SCROLLS — Per font, 75c.

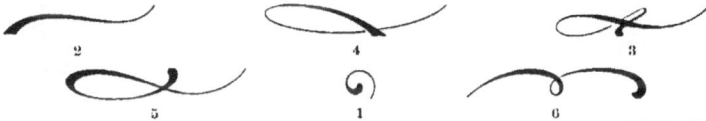

PACIFIC BIKES — Per font, $2.00; Single characters, 25c.

Inland Art Ornaments
Original

Single Ornaments may be ordered separately from any Series, at prices under the different characters.

INLAND ART ORNAMENTS, SERIES No. 1
PER FONT, 80c.

48001 — 20c.   48002 — 20c.   36001 — 15c.   48003 — 20c.   48004 — 25c.

INLAND ART ORNAMENTS, SERIES No. 2
PER FONT, 80c.

48005 — 20c.   48006 — 20c.   36002 — 20c.   48007 — 20c.   18008 — 20c.

INLAND ART ORNAMENTS, SERIES No. 3
PER FONT, 80c.

18009 — 20c.   48012 — 20c.   48011 — 20c.   48010 — 20c.   48013 — 20c.

INLAND ART ORNAMENTS, SERIES No. 4
PER FONT, 75c.

30002 — 10c.   24002 — 10c.   30004 — 10c.   30005 — 10c.

36003 — 15c.

Ornaments Nos. 30002A and 24005A are made
to face corresponding numbers here shown.
Price the same for each ornament,

24001 — 10c.   24005 — 10c.

Inland Art Ornaments

Original

INLAND ART ORNAMENTS, SERIES No. 5

PER FONT, 75c.

30006 — 10c.   24006 — 10c.   24009 — 10c.   24010 — 10c.   30003 — 12c.   24008 — 10c.

Ornaments Nos. 24008A and 30006A are made to face corresponding numbers above.
Price the same for each ornament.

INLAND ART ORNAMENTS, SERIES No. 6

PER FONT, 75c.

18013 — 6c.   18009 — 6c.   18029 — 8c, 18029A — 8c.   18008 — 6c.   18010 — 6c.

18034 — 6c.   18002 — 6c.   18030 — 8c.   18001 — 6c.   18032 — 6c.

INLAND ART ORNAMENTS, SERIES No. 7

PER FONT, 75c.

18038 — 8c.   18024 — 8c.   18039 — 8c.   18035 — 6c.   18037 — 8c.

18006 — 6c.   18012 — 6c.   18023 — 6c.   18036 — 6c.   18033 — 6c.   18005 — 6c.

INLAND ART ORNAMENTS, SERIES No. 8

PER FONT, 75c.

12025   12034   12029   12029A   12033   12030

12005   12001   12006

12035   12031   12026   12036   12032

Single Ornaments, 5c. each.

Wave Ornaments

⌒⌒

WAVE ORNAMENTS, SERIES No. 9

PER FONT, $1.50

24012 — 15c.

24013 — 10c.

18016 — 10c.

18018 — 10c.

36005 — 15c.

12014 — 6c.

36004 — 20c.

18014 — 10c.

24011 — 10c.

18017 — 8c.

WAVE ORNAMENTS, SERIES No. 10

PER FONT, $1.50

36007 — 20c.

18015 — 10c.

24017 — 10c.

18020 — 8c.

36008 — 15c.

12016 — 6c.

24015 — 15c.

24016 — 10c.

18019 — 10c.

18021 — 10c.

Inland Art Ornaments

Original

Single Ornaments may be ordered separately from any Series, at prices under the different characters.

INLAND ART ORNAMENTS, SERIES No. 11

PER FONT, 60c.

18040	18041	18042	18043	18044	18045	18046	18047

Single Ornaments, 7c. each.

INLAND ART ORNAMENTS, SERIES No. 12

PER FONT, 60c.

18048	18049	18050	18051	18052	12043	

Single Ornaments, 7c. each.

INLAND ART ORNAMENTS, SERIES No. 13

PER FONT, 75c.

24020	24022	24023	24018	24019	24024	24021

Single Ornaments, 9c. each.

INLAND ART ORNAMENTS, SERIES No. 14

PER FONT, 75c.

24025	24027	24028	24029	24030	24031	24026

Single Ornaments, 9c. each.

Inland Art Ornaments

Original

Single Ornaments may be ordered separately from any Series, at prices under the different characters.

INLAND ART ORNAMENTS, SERIES No. 15

PER FONT, 90c.

30008 — 10c.   30014 — 10c.   30011 — 10c.   30007 — 10c.   30010 — 10c.

30009 — 10c.   30012 — 10c.   30013 — 10c.

INLAND ART ORNAMENTS, SERIES No. 16

PER FONT, $1.00

36010 — 15c.   36012 — 15c.   36013 — 15c.   36014 — 15c.

36009 — 15c.   36011 — 15c.

INLAND ART ORNAMENTS, SERIES No. 18

PER FONT, 90c.

18004A — 25c.   48014 — 20c.   18016 — 20c.   48015 — 20c.   48009A — 20c.

Inland Art Ornaments

Original

Single Ornaments may be ordered separately from any Series, at prices under the different characters

INLAND ART ORNAMENTS, SERIES No. 19

PER FONT, 80c.

48005 — 20c.   48001A — 20c.   36001A — 15c.   18013A — 20c.   48005A — 20c.

INLAND ART ORNAMENTS, SERIES No. 20

PER FONT, 75c.

24005A — 10c.   24008A — 10c.   24008 — 10c.   24005 — 10c.

30006A — 10c.   30002 — 10c.   30002A — 10c.   30006 — 10c.

INLAND ART ORNAMENTS, SERIES No. 21

PER FONT, 75c.

18053 — 8c.   18053A — 8c.   12053 — 5c.   12053A — 5c.   18057 — 8c.   18057A — 8c.

12042 — 5c.   18051 — 8c.   18054A — 8c.   6040 — 3c

INLAND ART ORNAMENTS, SERIES No. 22

PER FONT, $1.00

48004 — 25c.   48009 — 20c. 48013A — 20c. 48013 — 20c.   48009A — 20c.   48004A — 25c.

𝔚reath 𝔒rnaments

Original

BUY THE
BEST

1897

BORDERS
MADE

WREATH ORNAMENTS, SERIES No. 17

Made in 8
sections which
will make various
combinations, all
justifying with Pt.
spaces and quads,
which are furnished
with every font.
Price, $1.00

Is cast with our
Superior Metal

NEW STYLE

Capable of Many
Combinations

Made in 7
pieces, which
will make var-
ious combina-
tions, justified
most readily
with pt. spaces
and quads

When

You

Get

Our

Type

SERIES No. 23

Per font, 75c.

You

Get

The

Very

Best

"New Art" Ornaments

Patent Pending

SERIES No. 24

ALL CHARACTERS CAST ON 24-POINT BODY

Fonts, measuring 30 inches, $2.50

2431   2430   2433   2432

2436   2429   2435   2429A   2434

Single characters of this series may be ordered separately, at the rate of 6 inches for 50c.

Our 2-Point Brass Rule No. 100 matches the line running through these Ornaments, and can readily be justified to join. The Rule running crosswise above is our 1-Point Rule No. 1.

Latest Inland Ornaments

Patent Pending

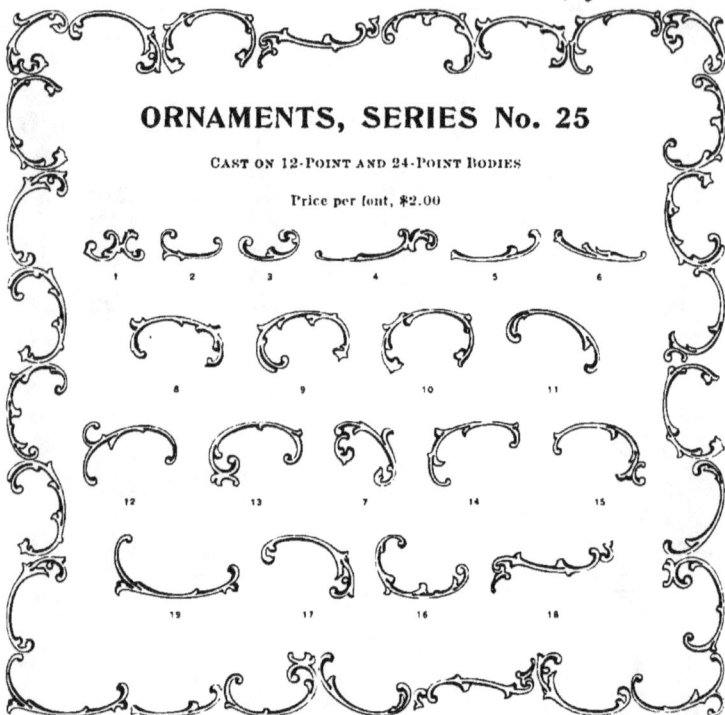

ORNAMENTS, SERIES No. 25

CAST ON 12-POINT AND 24-POINT BODIES

Price per font, $2.00

See page 80 for display of these Ornaments in connection with the Skinner Series.

INLAND ORNAMENTS, SERIES No. 26

CAST ON 12-POINT BODY

Price per font, $1.20

Single characters of Ornament Series Nos. 25 and 26 may be ordered separately.
12-Point body, 12 inches, 60c.; 24-Point body, 6 inches, 50c.

Newest Inland Ornaments

Original

Single Ornaments may be ordered separately from any Series, at prices under the different characters.

SERIES No. 27	SERIES No. 28	SERIES No. 29
Per font, 80c.	Per font, 80c.	Per font, 80c.

48017—20c.   48018—20c.

36017—15c.

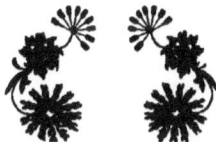

36018   36019
20c.   20c.

30015—20c.   30016—20c.

30017   30018
15c.   15c.

30019—15c.

36020
15c.   48019—20c.

36015—15c.   36016—15c.

42001—20c.

30020—15c.   30021—15c.

SERIES No. 30

Per font, 80c.

48021—25c.

36021—15c.

36022—15c.

48022—25c.

SERIES No. 31

Per font, $1.00

36023—25c.

42002—25c.

48020—25c.

54001—30c.

36024—25c.

Inland "Rugbys" and "Industrials"

Original

Single Ornaments may be ordered from either of these Series, at the price given. Fonts contain one of each.

ORNAMENTS, SERIES No. 32

EACH, 25C.   PER FONT, $2.00.

48023

60001

72001   72002   72003   72004   60002

72005   60003   72006

ORNAMENTS, SERIES No. 33

EACH, 25C.   PER FONT, $2.00.

60004   60005   60006   60007   60008

72007   72008   72009   72010   72011   72012

Kelmscott Ornaments
Original

Kelmscott Ornaments—In fonts of 6 inches each, at the prices given.

10-Point—30c.

12-Point—30c.

8-Point—30c.

24-Point—50c.

18-Point—40c.

14-Point—40c.

30-Point—50c.

36-Point—60c.

48-Point—75c.

Inland Holiday Cuts
Original

72101—30c.

72102—30c.

72103—30c.

72104—30c.

72105—30c.

72106—30c.

72107—30c.

72108—30c.

72109—30c.

72111—30c.

72112—30c.

72113—30c.

Inland Holiday Cuts

Original

  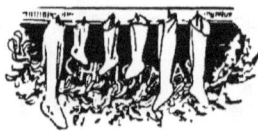 

48101 — 30c

72110 — 30c.   72114 — 30c.   72115 — 30c

72116 — 30c.   72117 — 30c.   72118 — 30c.   72119 — 30c.

72120 — 30c.   72121 — 30c.   72122 — 30c.   60101 — 30c.

108101* — 50c.   108102 — 50c.   108103* — 50c.

*Cuts marked * are electrotyped on wood or metal bases as desired.

Inland Holiday Cuts

Original

84101* — 50c.

84102* — 50c.

84103* — 50c.

90101* — 50c.

48102 — 25c.

90102* — 50c.

72123 — 30c.

120101* — 75c.

72124 — 30c.

18101 — 20c.

96101* — 60c.

78101 — 50c.

96102* — 60c.

* Cuts marked * are electrotyped on wood or metal bases as desired.

Inland Borders

Original

18-POINT BORDER NO. 1803; fonts of 24 inches, each, $1.00

12-POINT BORDER NO. 1203
Fonts of 24 inches, each, 75c.

INLAND BORDERS

Series No. 3

6-POINT BORDER NO. 603
Fonts of 48 inches, each, $1.25

24-POINT BORDER NO. 2403; fonts of 24 inches, each, $1.50

24-POINT BORDER NO. 2404; fonts of 24 inches, each, $1.50

12-POINT BORDER NO. 1204
Fonts of 24 inches, each, 75c.

INLAND BORDERS

Series No. 4

6-POINT BORDER NO. 604
Fonts of 48 inches, each, $1.25

18-POINT BORDER NO. 1804; fonts of 24 inches, each, $1.00

Inland Borders

Original

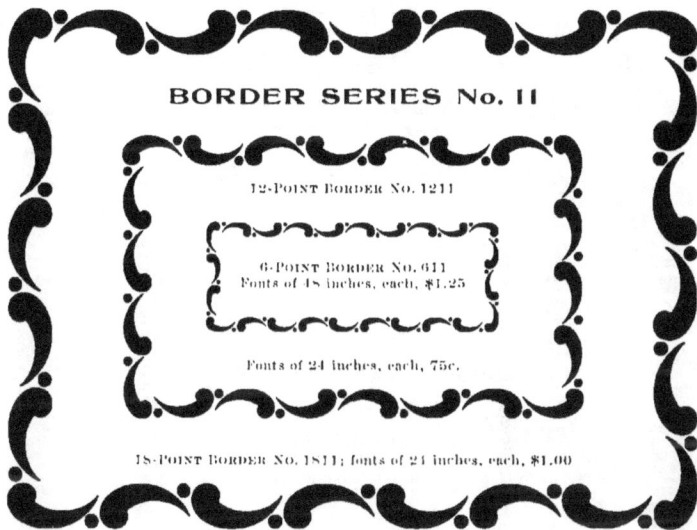

BORDER SERIES No. 11

12-POINT BORDER NO. 1211

6-POINT BORDER NO. 611
Fonts of 48 inches, each, $1.25

Fonts of 24 inches, each, 75c.

18-POINT BORDER NO. 1811; fonts of 24 inches, each, $1.00

These Borders fit exactly over one another for two-color work.

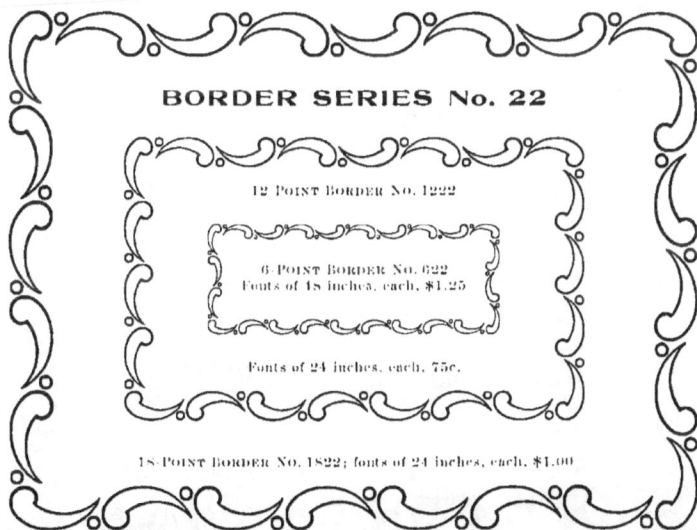

BORDER SERIES No. 22

12-POINT BORDER NO. 1222

6-POINT BORDER NO. 622
Fonts of 48 inches, each, $1.25

Fonts of 24 inches, each, 75c.

18-POINT BORDER NO. 1822; fonts of 24 inches, each, $1.00

Inland Borders

Original

Inland Borders

Series No. 27

Sizes and Prices

18-POINT BORDER No. 4827
Fonts of 12 inches, each, $1.50
36-POINT BORDER No. 3627
Fonts of 18 inches, each, $1.60
24-POINT BORDER No. 2427
Fonts of 21 inches, each, $1.60
18-POINT BORDER No. 1827
Fonts of 24 inches, each, $1.25
12 POINT BORDER No. 1227
Fonts of 24 inches, each, $1.00
6-POINT BORDER No. 627
Fonts of 36 inches, each, $1.20

The corresponding sizes of Border
Series Nos. 27 and 28 fit accurately over one another for
use in two-color work.

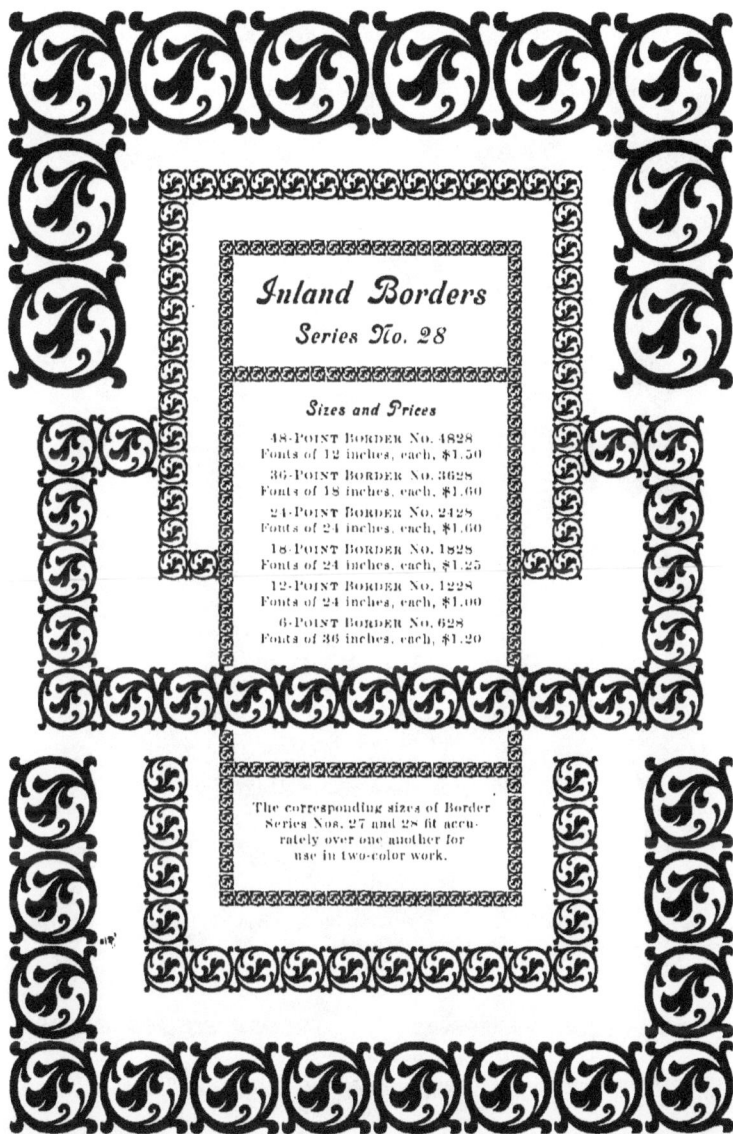

Inland Borders

Original

Inland Borders

Series No. 28

Sizes and Prices

48-Point Border No. 4828
Fonts of 12 inches, each, $1.50

36-Point Border No. 3628
Fonts of 18 inches, each, $1.60

24-Point Border No. 2428
Fonts of 24 inches, each, $1.60

18-Point Border No. 1828
Fonts of 24 inches, each, $1.25

12-Point Border No. 1228
Fonts of 24 inches, each, $1.00

6-Point Border No. 628
Fonts of 36 inches, each, $1.20

The corresponding sizes of Border
Series Nos. 27 and 28 fit accurately over one another for
use in two-color work.

Inland Borders

Original

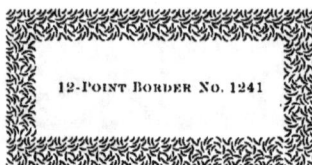

12-POINT BORDER No. 1241

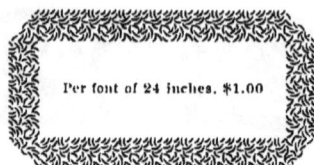

Per font of 24 inches, $1.00

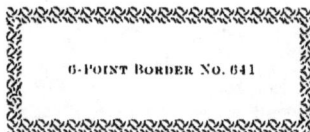

6-POINT BORDER No. 641

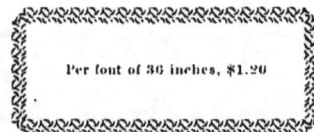

Per font of 36 inches, $1.20

Inland Borders

Original

12-POINT BORDER NO. 1247; fonts of 24 inches, each, $1.00

6-POINT BORDER NO. 647

Inland Borders

Series No. 47

Fonts of 36 inches, each, $1.20

18-POINT BORDER NO. 1847; fonts of 24 inches, each, $1.25

12-POINT BORDER NO. 1248; fonts of 24 inches, each, $1.00

6-POINT BORDER NO. 648

Inland Borders

Series No. 48

Fonts of 36 inches, each, $1.20

18-POINT BORDER NO. 1848; fonts of 24 inches, each, $1.25

Inland Borders

Original

24-POINT BORDER NO. 2466; fonts of 24 inches, each, $1.60

12-POINT BORDER NO. 1266
Fonts of 24 inches, each, $1.00

INLAND BORDERS
Series No. 66

6-POINT BORDER NO. 666
Fonts of 36 inches, each, $1.20

18-POINT BORDER NO. 1866; fonts of 24 inches, each, $1.25

18-POINT BORDER NO. 1867; fonts of 24 inches, each, $1.25

12-POINT BORDER NO. 1267
Fonts of 24 inches, each, $1.00

INLAND BORDERS
Series No. 67

6-POINT BORDER NO. 667
Fonts of 36 inches, each, $1.20

24-POINT BORDER NO. 2467; fonts of 24 inches, each, $1.60

Inland Borders

Original

12-POINT BORDER NO. 1268; fonts of 24 inches, each, $1.00

6-POINT BORDER NO. 668

INLAND BORDERS
Series No. 68

Fonts of 36 inches, each, $1.20

18-POINT BORDER NO. 1868; fonts of 24 inches, each, $1.25

24-POINT BORDER NO. 2407; fonts of 24 inches, each, $1.50

12-POINT BORDER NO. 1207

INLAND BORDERS
Series No. 7

Fonts of 24 inches, each, 75c.

18-POINT BORDER NO. 1807; fonts of 24 inches, each, $1.00

Inland Borders

Original

12-POINT BORDER NO. 1262; fonts of 24 inches, each, $1.00

6-POINT BORDER NO. 662

INLAND BORDERS
Series No. 62

Fonts of 36 inches, each, $1.20

18-POINT BORDER NO. 1862; fonts of 24 inches, each, $1.25

These Borders are cast to fit exactly over one another for two color work.

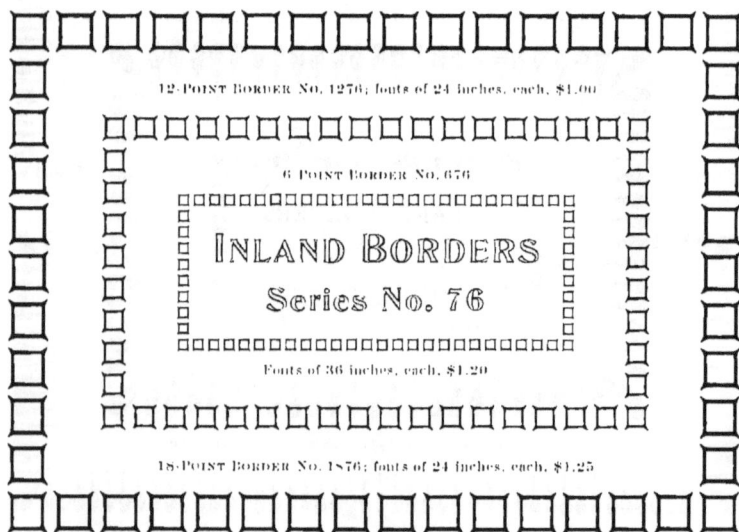

12-POINT BORDER NO. 1276; fonts of 24 inches, each, $1.00

6-POINT BORDER NO. 676

INLAND BORDERS
Series No. 76

Fonts of 36 inches, each, $1.20

18-POINT BORDER NO. 1876; fonts of 24 inches, each, $1.25

Inland Borders

Original

12-POINT BORDER NO. 1263; fonts of 24 inches, each, $1.00

6-POINT BORDER NO. 663

INLAND BORDERS
Series No. 63

Fonts of 36 inches, each, $1.20

18-POINT BORDER NO. 1863; fonts of 24 inches, each, $1.25

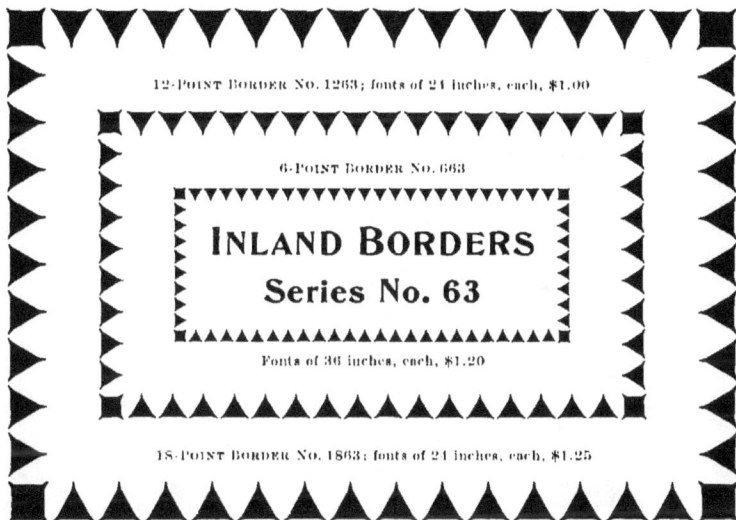

These Borders are cast to fit exactly over one another for two-color work.

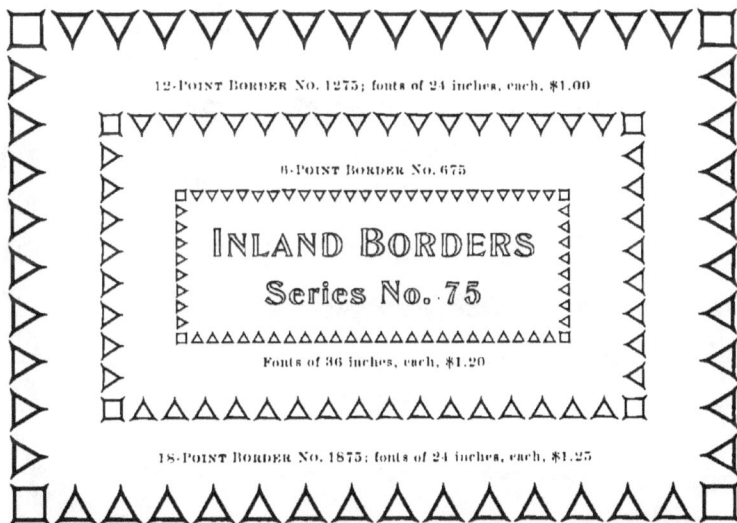

12-POINT BORDER NO. 1275; fonts of 24 inches, each, $1.00

6-POINT BORDER NO. 675

INLAND BORDERS
Series No. 75

Fonts of 36 inches, each, $1.20

18-POINT BORDER NO. 1875; fonts of 24 inches, each, $1.25

Inland Borders

Original

BORDER SERIES No. 84

18-POINT BORDER NO. 1884
Fonts of 36 inches, each, $1.90
12-POINT BORDER NO. 1284
Fonts of 30 inches, each, $1.50
6-POINT BORDER NO. 684
Fonts of 36 inches, each, $1.50

84

Following Characters are cut for each size:

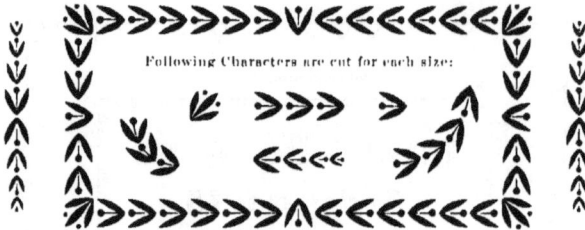

Series Nos. 84 and 86 fit over one another for two-color work.
The round corners of each size of this series, excepting the smaller
of the 6-Point, are cast on mortised bodies.

Inland Borders

Original

BORDER SERIES No. 86

18-POINT BORDER NO. 1886
Fonts of 36 inches, each, $1.90

12-POINT BORDER NO. 1286
Fonts of 30 inches, each, $1.50

6-POINT BORDER NO. 686
Fonts of 36 inches, each, $1.50

Following characters are cut for each size:

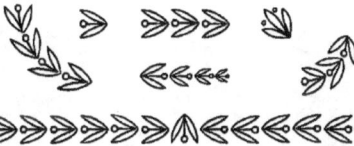

Series Nos. 84 and 86 fit over one another for two-color work.
The round corners of each size of this series, excepting the smaller
of the 6-Point, are cast on mortised bodies.

Inland Borders

Original

INLAND BORDERS

SERIES No. 78

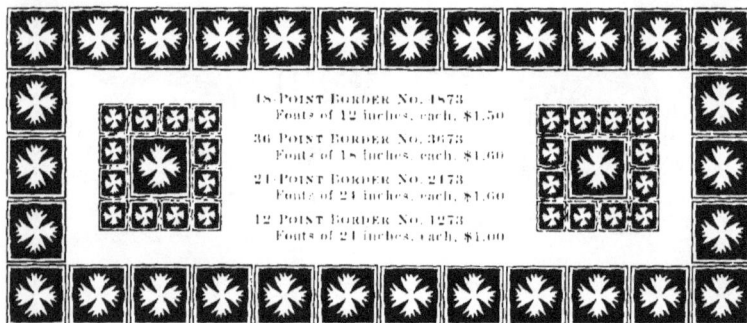

18 POINT BORDER No. 1873
Fonts of 12 inches, each, $1.50

36 POINT BORDER No. 3673
Fonts of 18 inches, each, $1.60

24 POINT BORDER No. 2473
Fonts of 24 inches, each, $1.60

12 POINT BORDER No. 1273
Fonts of 24 inches, each, $1.00

These Borders are very suitable for making
up into forms for tint backgrounds

Inland Borders

Original

6-POINT BORDERS

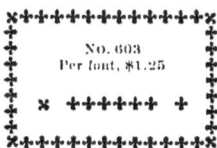

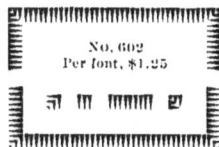

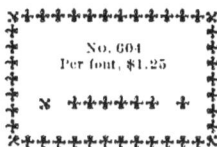

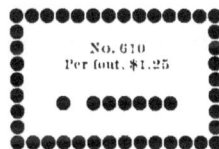

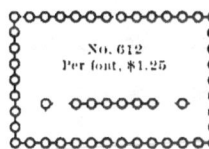

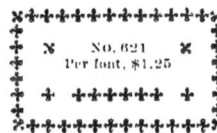

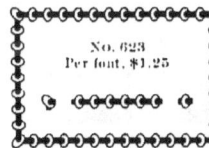

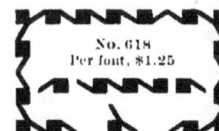

No. 601
Per font, $1.25

No. 603
Per font, $1.25

No. 608
Per font, $1.25

No. 602
Per font, $1.25

No. 604
Per font, $1.25

No. 609
Per font, $1.25

No. 610
Per font, $1.25

No. 611
Per font, $1.25

The right and left
characters have
different nicks.

No. 612
Per font, $1.25

No. 614
Per font, $1.25

No. 615
Per font, $1.25

No. 616
Per font, $1.25

No. 617
Per font, $1.25

No. 621
Per font, $1.25

No. 623
Per font, $1.25

No. 618
Per font, $1.25

No. 622
Per font, $1.25

The right and left
characters have
different nicks.

No. 640
Per font, $1.25

Inland Borders

Original

6-POINT BORDERS

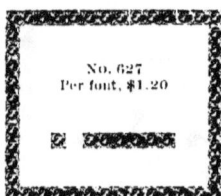

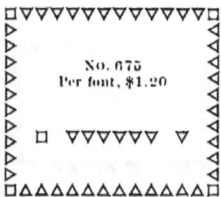

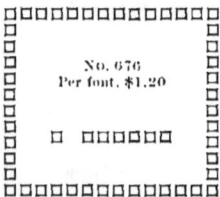

No. 627
Per font, $1.20

No. 628
Per font, $1.20

No. 641
Per font, $1.20

No. 647
Per font, $1.20

No. 648
Per font, $1.20

No. 658
Per font, $1.20

No. 661
Per font, $1.20

No. 662
Per font, $1.20

No. 663
Per font, $1.20

No. 666
Per font, $1.20

No. 667
Per font, $1.20

No. 668
Per font, $1.20

No. 675
Per font, $1.20

No. 676
Per font, $1.20

No. 678
Per font, $1.20

Inland Borders

Original

6-POINT BORDERS

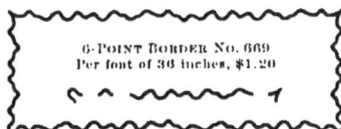

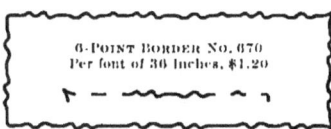

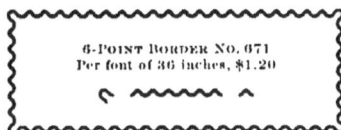

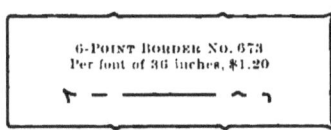

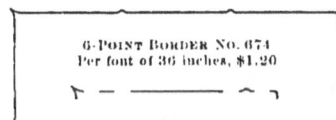

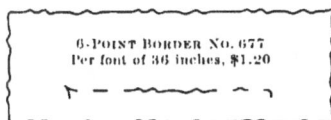

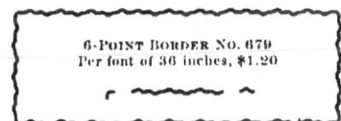

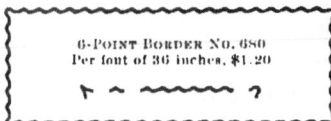

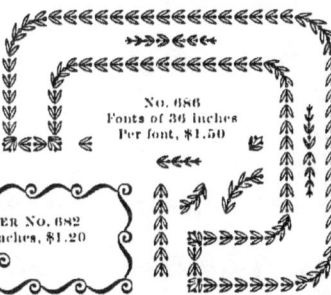

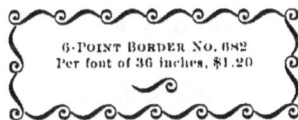

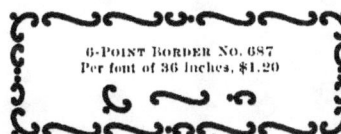

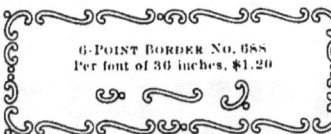

6-POINT BORDER No. 669
Per font of 36 inches, $1.20

6-POINT BORDER No. 670
Per font of 36 inches, $1.20

6-POINT BORDER No. 671
Per font of 36 inches, $1.20

6-POINT BORDER No. 673
Per font of 36 inches, $1.20

6-POINT BORDER No. 674
Per font of 36 inches, $1.20

6-POINT BORDER No. 677
Per font of 36 inches, $1.20

6-POINT BORDER No. 679
Per font of 36 inches, $1.20

6-POINT BORDER No. 680
Per font of 36 inches, $1.20

No. 684
Fonts of 36 inches
Per font, $1.50

No. 686
Fonts of 36 inches
Per font, $1.50

6-POINT BORDER No. 682
Per font of 36 inches, $1.20

6-POINT BORDER No. 687
Per font of 36 inches, $1.20

6-POINT BORDER No. 688
Per font of 36 inches, $1.20

Inland Borders

Original

12-POINT BORDERS

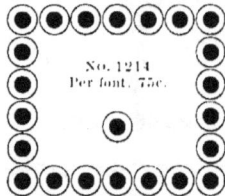

12-POINT BORDER No. 1202
Per font, 75c.

No. 1203
Per font, 75c.

No. 1207
Per font, 75c.

No. 1204
Per font, 75c.

No. 1209
Per font, 75c.

No. 1208
Per font, 75c.

No. 1210
Per font, 75c.

No. 1211
Per font, 75c.

These characters
have different nicks.

No. 1214
Per font, 75c.

No. 1213
Per font, 75c.

No. 1215
Per font, 75c.

No. 1212
Per font, 75c.

No. 1216
Per font, 75c.

Inland Borders

Original

12-POINT BORDERS

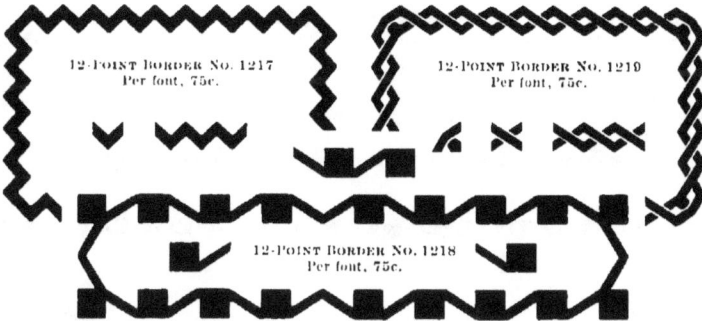

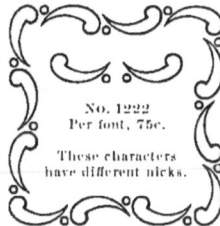

12-POINT BORDER No. 1217
Per font, 75c.

12-POINT BORDER No. 1219
Per font, 75c.

12-POINT BORDER No. 1218
Per font, 75c.

No. 1220
Per font, 75c.

No. 1221
Per font, 75c.

No. 1222
Per font, 75c.

These characters
have different nicks.

No. 1223
Per font, 75c.

No. 1224
Per font, 75c.

No. 1225
Per font, 75c.

12-POINT BORDER No. 1226
Per font, 75c.

No. 1227 — Per font, $1.00

No. 1228 — Per font, $1.00

Inland Borders

Original

12-POINT BORDERS

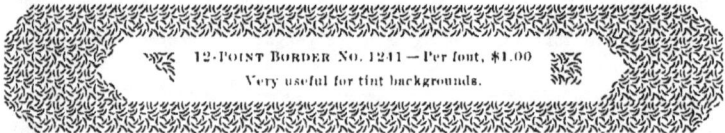

12-POINT BORDER NO. 1229 — Per font, 75c.

12-POINT BORDER NO. 1230 — Per font, 75c.

12-POINT BORDER NO. 1231 — Per font, 75c.

12-POINT BORDER NO. 1232 — Per font, 75c.

12-POINT BORDER NO. 1233
Per font, 75c.

12-POINT BORDER NO. 1234
Per font, 75c.

12-POINT BORDER NO. 1235 — Per font, 75c.

12-POINT BORDER NO. 1236 — Per font, 75c.

12-POINT BORDER NO. 1241 — Per font, $1.00
Very useful for tint backgrounds.

12-POINT BORDER NO. 1242 — Per font, 75c.

12-POINT BORDER NO. 1244 — Per font, 75c.

Inland Borders

Original

12-POINT BORDERS

12-POINT BORDER No. 1247
Per font, $1.00

12-POINT BORDER No. 1248
Per font, $1.00

12-POINT BORDER No. 1249
Per font, $1.00

12-POINT BORDER No. 1250
Per font, $1.00

12-POINT BORDER No. 1251
Per font, $1.00

12-POINT BORDER No. 1252
Per font, $1.00

12-POINT BORDER No. 1253 — Per font, $1.00

The right and left characters of Border No. 1253 have different nicks.

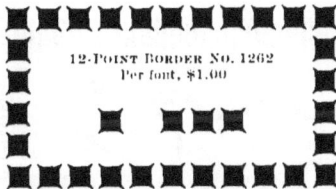

12-POINT BORDER No. 1262
Per font, $1.00

12-POINT BORDER No. 1263
Per font, $1.00

Inland Borders

Original

12-POINT BORDERS

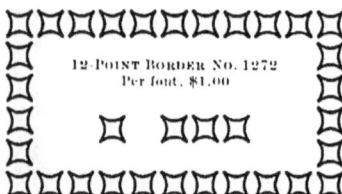

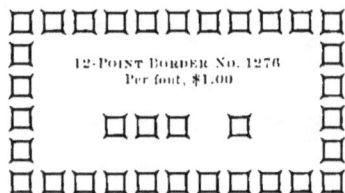

12-POINT BORDER No. 1266 — Per font, $1.00

The right and left characters of Border No. 1266 have different nicks.

12-POINT BORDER No. 1267
Per font, $1.00

12-POINT BORDER No. 1268
Per font, $1.00

12-POINT BORDER No. 1272
Per font, $1.00

12-POINT BORDER No. 1273
Per font, $1.00

12-POINT BORDER No. 1275
Per font, $1.00

12-POINT BORDER No. 1276
Per font, $1.00

12-POINT BORDER No. 1278 — Per font, $1.00

12-POINT BORDER No. 1282
Per font, $1.00

Inland Borders

Original

12-POINT BORDERS

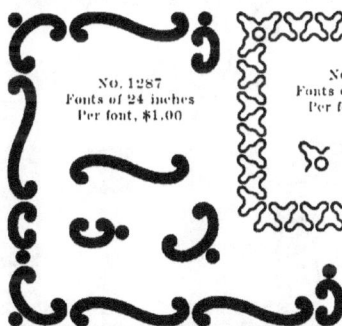

No. 1281
Fonts of 30 inches
Per font, $1.25

No. 1289
Fonts of 30 inches
Per font, $1.25

12-POINT BORDER No. 1284
Fonts of 30 inches
Per font, $1.50

12-POINT BORDER No. 1286
Fonts of 30 inches
Per font, $1.50

No. 1287
Fonts of 24 inches
Per font, $1.00

No. 1285
Fonts of 24 inches
Per font, $1.00

No. 1288
Fonts of 24 inches
Per font, $1.00

Excepting No. 1285, the above Borders fit exactly over one another for two-color work.

Inland Borders

Original

18-POINT BORDERS

18-POINT BORDER No. 1803
Per font, $1.00

18-POINT BORDER No. 1804
Per font, $1.00

18-POINT BORDER No. 1811
Per font, $1.00

The right and left characters
are nicked differently.

18-POINT
BORDER
No. 1807

18-POINT
BORDER
No. 1819

Borders Nos. 1811 and 1822 fit exactly over
one another for two-color work.

Per font, $1.00

Per font, $1.00

18-POINT BORDER No. 1822
Per font, $1.00

The right and left characters
are nicked differently.

18-POINT BORDER No. 1827
Per font, $1.25

18-POINT BORDER No. 1828
Per font, $1.25

Borders Nos. 1803 and 1804, and 1827 and 1828, fit exactly over one another for two-color work.

Inland Borders

Original

Fonts of 24 inches each

18-POINT BORDERS

18-POINT BORDER No. 1829 — Per font, $1.00

18-POINT BORDER No. 1830 — Per font, $1.00

The right and left characters of Border No. 1829 are nicked differently.

18-POINT BORDER No. 1832 — Per font, $1.00

18-POINT BORDER No. 1833 — Per font, $1.00

18-POINT BORDER No. 1834 — Per font, $1.00

18-POINT BORDER No. 1835 — Per font, $1.00

18-POINT BORDER No. 1836 — Per font, $1.00

18-POINT BORDER No. 1837 — Per font, $1.00

Inland Borders

Original

18-POINT BORDERS

18-POINT BORDER No. 1838 — Per font, $1.00

18-POINT BORDER No. 1839 — Per font, $1.00

18-POINT BORDER No. 1844 — Per font, $1.00

18-POINT BORDER No. 1853 — Per font, $1.25

The right and left characters of Borders Nos. 1853 and 1854 are nicked differently.

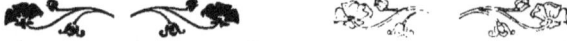

These Borders fit exactly over one another for two color work.

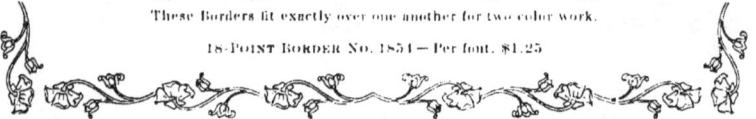

18-POINT BORDER No. 1854 — Per font, $1.25

18-POINT BORDER No. 1855 — Per font, $1.00

18-POINT BORDER No. 1859 — Per font, $1.00

18-POINT BORDER No. 1860 — Per font, $1.00

Inland Borders

Original

18-POINT BORDERS

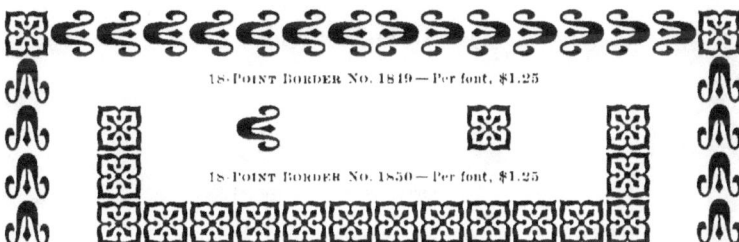

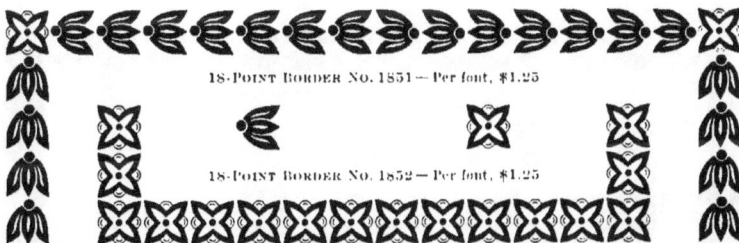

18-POINT BORDER No. 1845 — Per font, $1.00

18-POINT BORDER No. 1846 — Per font, $1.00

18-POINT BORDER No. 1847 — Per font, $1.25

18-POINT BORDER No. 1848 — Per font, $1.25

18-POINT BORDER No. 1849 — Per font, $1.25

18-POINT BORDER No. 1850 — Per font, $1.25

18-POINT BORDER No. 1851 — Per font, $1.25

18-POINT BORDER No. 1852 — Per font, $1.25

Inland Borders

Original

18-POINT BORDERS

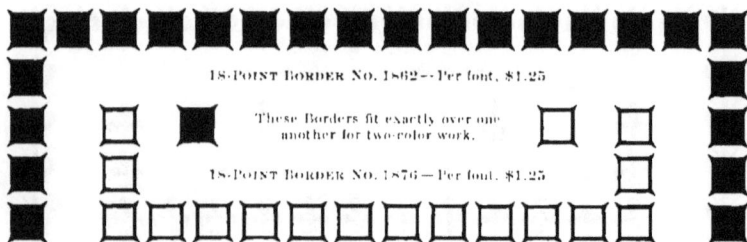

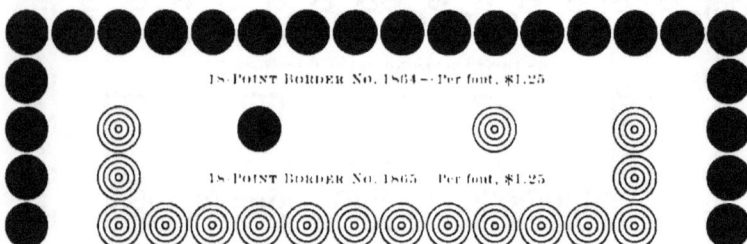

18-POINT BORDER No. 1862—Per font, $1.25

These Borders fit exactly over one
another for two-color work.

18-POINT BORDER No. 1876—Per font, $1.25

18-POINT BORDER No. 1863—Per font, $1.25

These Borders fit
over one another for
two-color work.

18-POINT BORDER No. 1875—Per font, $1.25

18-POINT BORDER No. 1864—Per font, $1.25

18-POINT BORDER No. 1865—Per font, $1.25

18-POINT BORDER No. 1868—Per font, $1.25

18-POINT BORDER No. 1883—Per font, $1.25

Inland Borders

Original

18-POINT BORDERS

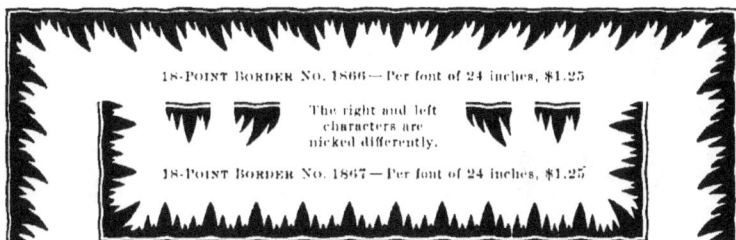

18-POINT BORDER No. 1866 — Per font of 24 inches, $1.25

The right and left
characters are
nicked differently.

18-POINT BORDER No. 1867 — Per font of 24 inches, $1.25

18-POINT BORDER No. 1886
Per font of 36 inches, $1.90

These Borders fit
over one another for
two-color work.
—
The round corners
of each Border are
on mortised bodies.

18-POINT BORDER No. 1884
Per font of 36 inches, $1.90

Inland Borders

Original

Fonts of 24 inches each

24-POINT BORDERS

24-Point Border No. 2403
Per font, $1.50

24-Point Border No. 2404
Per font, $1.50

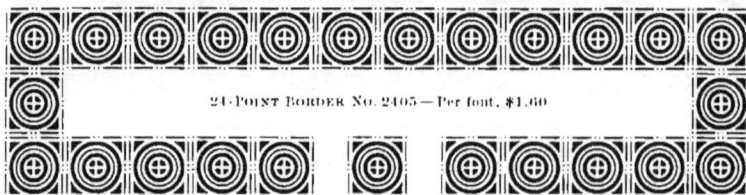

24-Point Border No. 2405 — Per font, $1.60

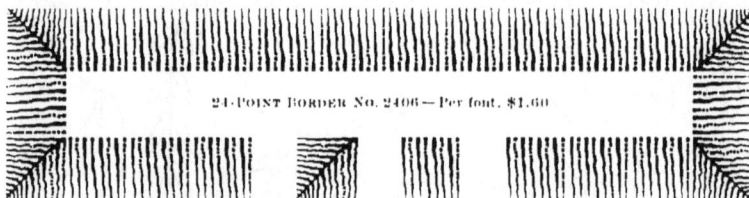

24-Point Border No. 2406 — Per font, $1.60

24-Point Border No. 2407 — Per font, $1.50

24-Point Border No. 2474 — Per font, $1.60

Inland Borders

Original

24-POINT BORDERS

24-POINT BORDER No. 2427 — Per font, $1.60

24-POINT BORDER No 2428 — Per font, $1.60

24-POINT BORDER No. 2466 — Per font, $1.60

24-POINT BORDER No 2467 — Per font, $1.60

24 POINT BORDER No. 2473 — Per font, $1 60

Inland Band Borders

Fonts of 12 inches **24-POINT BAND BORDER** Per font, $0.80

No. 2401

Fonts of 12 inches, each **30-POINT BAND BORDERS** Per font, each, $0.95

No. 3003

No. 3015

No. 3016

No. 3017

No. 3018

No. 3019

No. 3020

No. 3021

No. 3022

Inland Band Borders

Fonts of 12 inches, each ## 30-POINT BAND BORDERS Per font, each, $0.95

No. 3023

Fonts of 12 inches, each ## 36-POINT BAND BORDERS Per font, each, $1.10

No. 3603

No. 3604

No. 3605

No. 3609

No. 3617

No. 3618

Fonts of 12 inches, each ## 42-POINT BAND BORDERS Per font, each, $1.25

No. 4205

Inland Band Borders

No. 4210

No. 4211

No. 4215

No. 4216

No. 4219

No. 4220

No. 4221

No. 4222

Inland Band Borders

Fonts of 12 inches, each   **42-POINT BAND BORDERS**   Per font, each, $1.25

No. 4223

Fonts of 12 inches, each   **48-POINT BAND BORDERS**   Per font, each, $1.40

No. 4801

No. 4809

Fonts of 12 inches   **54-POINT BAND BORDER**   Per font, $1.55

No. 5104

Fonts of 12 inches   **60-POINT BAND BORDER**   Per font, $1.70

No. 6011

Fonts of 12 inches   **72-POINT BAND BORDER**   Per font, $2.00

No. 7201

MAILING-LIST MATERIAL

10-Point Mailing-List Type 18c. per pound

W H Woodward 1⅜99 ST LOUIS MO	Th L DeVinne 2⅝98 NEW YORK N Y	Inland Printer x CHICAGO ILL
JohnAThayer 23⅜97 PHILADELPHIA PA	KeystonePress5⅝96 WELLSTON OHIO	T W Thomas 31⅜98 TOLEDO OHIO
W H Wright 26⅜02 PITTSBURG PA	HolbrookPtgCo3⅜97 NEWARK N J	F J Pierson 15⅝04 FLINT MICH
Thad B Mead 30⅞98 NEW YORK N Y	Henry R Boss 8⅜99 CHICAGO ILL	Little&Becker7⅝97 ST LOUIS MO
E D Wesoott 20⅝97 READING PA	Alb W Dennis 7⅝97 LYNN MASS	D B Landis 12⅜98 LANCASTER PA
CTHenderson 23⅜01 TOULON ILL	W H Bevis 8⅜02 PAWTUCKET R I	E L Wepf 21⅜01 DENVER COLO
Press & Printer x BOSTON MASS	W A Donnelly 7⅞98 ROCHESTER N Y	LGraham&Son 31⅜98 NEW ORLEANS LA
DorseyPtgCo 23⅝97 DALLAS TEX	Chas Collier 2⅜98 SHREVE O	Gazette x ST JOSEPH MO

Every character of this face is cast on en-set, and as the spacing is done solely with en quads, it may be readily seen that mailing-lists are set up very quickly and changes and corrections easily made with this type. It runs no wider than ordinary 10-Point Romans. Measure of composing sticks should be a multiple of 10-Point.

MAILING-LIST LOGOTYPES

Supplied in any quantity, and assorted according to purchaser's desire, at the prices given.

10-Point En-Set Mailing-List Logotypes   18c. per pound

Jan Feb Mar Apr May Jun Jul Aug Sep Oct Nov Dec

9-Point En-Set Mailing-List Logotypes   50c. per pound

Jan Feb Mar Apr May Jun Jul Aug Sep Oct Nov Dec

9-Point Em-Set Mailing-List Logotypes   50c. per pound

Jan Feb Mar Apr May Jun Jul Aug Sep Oct Nov Dec

8-Point Em-Set Mailing-List Logotypes   53c. per pound

Jan Feb Mar Apr May Jun Jul Aug Sep Oct Nov Dec

We also supply other fonts assorted to order for mailing-lists. Mailing galleys, slugs and reglets supplied on short notice. We are agents for all makes of mailing-machines, and will furnish them at the manufacturer's terms.

TIME-TABLE FIGURES

Being cast on Standard Line, these Figures will line with all our Roman and Job faces cast on the same bodies, consequently the printer is enabled to use any one or more faces he may select for station names or other reading matter. All our styles of Leaders also line with these figures.

Light-Face Figures are ordinarily used for A. M., and the Heavy-Face for P. M. time.

Supplied in any quantity, and assorted according to purchaser's desire, at the prices given.

	Per pound
8-POINT SPECIAL LIGHT-FACE TIME-TABLE FIGURES — EN SET	$1.00

1 2 3 4 5 6 7 8 9 0

8-POINT SPECIAL HEAVY-FACE TIME-TABLE FIGURES — EN SET	$1.00

1 2 3 4 5 6 7 8 9 0

6-POINT SPECIAL LIGHT-FACE TIME-TABLE FIGURES — EN SET	$1.28

1 2 3 4 5 6 7 8 9 0

6-POINT SPECIAL HEAVY-FACE TIME-TABLE FIGURES — EN SET	$1.26

1 2 3 4 5 6 7 8 9 0

5½-POINT SPECIAL LIGHT-FACE TIME-TABLE FIGURES — EN SET	$1.60

1 2 3 4 5 6 7 8 9 0

5½-POINT SPECIAL HEAVY-FACE TIME-TABLE FIGURES — EN SET	$1.60

1 2 3 4 5 6 7 8 9 0

5-POINT SPECIAL LIGHT-FACE TIME-TABLE FIGURES — EN SET	$2.00

1 2 3 4 5 6 7 8 9 0

5-POINT SPECIAL HEAVY-FACE TIME-TABLE FIGURES — EN SET	$2.00

1 2 3 4 5 6 7 8 9 0

6-POINT FRENCH CLARENDON FIGURES — EN SET	$1.28

$ 1 2 3 4 5 6 7 8 9 0

SPECIMEN OF 6-POINT SPECIAL TIME-TABLE FIGURES IN USE.

13 Ex. Sun	19 Ex. Sun	23 Daily	31 Daily	37 Daily	STATIONS	14 Ex. Sun	48 Sunday	22 Ex. Sun	34 Daily	38 Ex. Mon
6 45	11 50	3 15	6 30	11 45	Lv. St. Louis U. D. Ar.	6 30	10 30	12 35	8 20	10 40
6 52	11 57	3 22	6 37	11 52	......Grand Avenue......	6 21	10 21	12 26	8 11	10 31
6 55	11 59	3 25	6 40	11 55	Vandeventer Avenue	6 18	10 18	12 23	8 08	10 28
6 58	12 03	3 28	6 43	11 58	...... Tower Grove	6 15	10 15	12 20 *	8 05	*10 25
7 01	12 06	3 31	6 46	12 01	.........Howard's.........	6 09	10 11	12 16 *	8 01	*10 21
7 03	12 08	3 33	6 47	12 03	........Cheltenham........	6 07	10 09	12 14 *	7 59	*10 19
7 05	12 10	3 35	6 48	12 05	.... Clifton Heights	6 05	10 07	12 12 *	7 57	*10 17
7 07	12 12	3 37	6 51	12 07	............Benton..........	6 03	10 05	12 10	7 55	10 15
7 09	12 14	3 39	6 53	12 09	..........Ellendale........	6 00	10 02	12 07 *	7 53	*10 13
* 7 11	*12 16	3 41	6 55	*12 11	.... Maplewood	* 5 58	10 00	12 05 *	7 51	*10 12
* 7 12	*12 18	3 43	6 57	*12 12	.... Sutton	* 5 57	9 58	*12 03 *	7 49	*10 10
* 7 13	*12 19	3 44	6 59	*12 13	........Edgebrook........	5 55	9 56	*12 01 *	7 47	*10 09
* 7 14	*12 20	3 45	7 01	*12 14	......Lake Junction......	5 54	9 55	*11 59 *	7 45	*10 08
* 7 16	12 22	3 47	7 04	*12 16	........Tuxedo Park........	5 52	9 53	11 57 *	7 43	*10 06
7 20	12 25	3 50	7 08	12 20	.... Webster	5 50	9 50	11 55	7 40	10 04
* 7 23	12 28	3 53	7 12	*12 23	..........Glendale..........	* 5 46	9 46	*11 51 *	7 36	*10 00
* 7 25	*12 30	3 55	7 15	*12 25	..........Oakland..........	* 5 44	9 44	*11 49 *	7 34	* 9 58
7 27	12 32	3 57	7 17	*12 27	.........Woodlawn.........	* 5 42	9 42	11 47 *	7 32	* 9 57
7 30	12 35	4 00	7 20	12 30	Ar.....Kirkwood....Lv.	5 40	9 40	11 45	7 30	9 55

Being cast on Point-sets and to multiples of spaces, the figures of every one of our Roman and Job faces are available with easy justification for all classes of tabular matter.

POINT-SET FIGURES

For Time-Tables, Tariffs, Etc.

The following specimens of tabular matter show the superiority of our system of casting Figures
and Punctuation Marks on Point Sets, or widths, and also give the printer a large variety from which
to select.

6-POINT FIGURES, VARIOUS SETS

1	2	3	4		5	6	7	8
7.50AM	8.15PM	8.40AM	8.30PM Lv. St. Louis Ar.	11.08	2.10	10.40	11.30
8.27	9.00	9.17	9.05 ..... Barracks	10.30	1.32	10.00	10.55
8.51	9.15	9.41	9.27 ..... Kimswick	10.04	1.06	9.35	10.36
* 8.57		* 9.46	.Sulphur Springs.	9.58	12.58	* 9.38	....
9.13	9.38	9.56	9.39 .... Pevely........	9.43	12.42	9.25	10.15
9.51	9.49	10.28	10.05 Ar... De Soto... Lv.	9.05	12.04PM	8.36	10.00
9.56	8.54	10.32	10.10 Lv. De Soto. Ar.	9.00	11.58	8.28	9.56
*10.08	10.23	10.44	*10.22 ...... Vineland	8.48	11.45	8.16	* 9.44
10.17	10.32	10.52	10.29 ..... Blackwell	8.39	11.36	8.10	9.35
10.41	11.00	11.13	10.53 ... Mineral Point...	8.18	11.15	7.48	9.12
11.15	11.35	11.45	11.20 Ar.. Bismarck .. Lv.	7.45PM	10.40AM	7.25PM	8.50AM

1. ROMAN No. 23
2. ITALIC No. 23
3. OLD STYLE No. 9 } Figures 3-Point set,
4. OLD STYLE ITALIC No. 9 } Period
5. CONDENSED GOTHIC No. 1 } 2-Point set.

6. CONDENSED WOODWARD,
   Figures 2½-Point set, Period 2-Point set.
7. LATIN, Figures 4-Point, Period 3-Point set.
8. BRUCE TITLE No. 2.
   Figures 4½-Point set, Period 2-Point set.

6-POINT FIGURES, 4-POINT SET

	9	10	11	12	13	14	15	16
St. Louis...Le.	1.40AM	2.10AM	1.05AM	1.55AM	12.55PM	1.45PM	12.25PM	12.40PM
Benton.........	1.53	2.23	1.18	2.04	1.00	2.00	12.38	1.02
Kirkwood .....:	2.14	2.44	2.35	2.26	1.28	2.18	1.20	1.30
Eureka	* 3.06	* 3.32	2.56	3.40	* 2.18	*	1.56	2.22
Pacific	4.17	4.42	3.54	4.35	3.33	4.22	2.40	3.35
Washington...	4.38	5.00	4.00	4.48	3.40	4.48	3.58	3.46
Chamois.....	6.25	6.48	5.50	6.36	5.36	6.28	4.05	3.40
Jefferson City	8.20	8.35	7.42	8.09	7.25	8.24	5.21	7.32
Tipton	* 9.36	*	9.16	9.42	8.57	9.40	7.00	⅜ ......
Sedalia........	10.52	10.44	10.09	10.58	10.00	10.58	8.32	10.03
Holden ..... Ar.	12.00	11.56	11.35	12.00	11.26	11.55	10.50	11.30

9. ROMAN No. 20, Period 2-Point set.
   3-Point set Period cast to order if wanted.
10. FRENCH OLD STYLE, Period 2-Point set.
    3-Point set Period cast to order if wanted.
11. CONDENSED No. 2, Period 3-Point set.

12. GOTHIC No. 6, Period 3-Point set.
13. FULL-FACE No. 1, Period 3-Point set.
14. CONDENSED No. 1, Period 3-Point set.
15. GOTHIC ITALIC No. 1, Period 3-Point set.
16. TUDOR BLACK, Period 3-Point set.

6-POINT FIGURES, 5-POINT SET

	17	18	19	20	21	22	23
Chicago .... $	4.35	$ 8.75	$10.25	$ 8.50	$ 6.00	$ 9.75	$12.60
Peoria	5.60	9.05	10.95	10.25	6.50	10.50	13.80
Springfield ...	6.10	9.60	13.60	13.75	7.25	12.50	15.00
Alton	8.15	10.25	15.75	14.30	9.30	15.60	18.35
Saint Louis...	9.20	10.90		15.95	10.75	18.25	20.75
Pacific		22.00	20.35	18.50	12.00	20.45	25.40
Sedalia	14.75	26.35	31.40	29.95	15.00	28.30	31.30
Kansas City....	16.00	30.45	35.30	32.25	20.65	32.75	
Topeka	28.70	42.30	46.80	44.10	33.40	43.15	53.65
Salina	36.25	56.10	60.25	53.00	41.35	50.00	68.45
Denver	48.10	62.40	75.00	60.80	54.50	68.40	79.00

17. HALF-TITLE, Period 3-Point set.
18. GOTHIC No. 1, Figure 1 cast either 3-Point
    or 5-Point set, Period 3-Point set.
19. SKINNER, Period 2-Point set.
    3-Point set Period cast to order if wanted.
20. ANTIQUE No. 1, Period 3-Point set.

21. WOODWARD, Figure 1 cast either 3-Point or
    5-Point set, Period 3-Point set.
22. EDWARDS, Period 2½-Point set.
    3-Point set Period cast to order if wanted
23. IONIC, Period 3-Point set.
    DO NOT ORDER BY THESE INDEX NUMBERS.

POINT-SET FIGURES

For Time-Tables, Tariffs, Etc.

6-POINT FIGURES, 6-POINT SET, AND MIXED DISPLAY

	24	25	26	27	28	29
Chicago	$ 5.75	$10.50	$ 8.80	$12.30	$11.50	$12.60
Peoria	8.00	15.75	12.75	16.60	18.90	20.35
Springfield	10.50	20.35	18.20	20.75	23.50	25.75
Alton	16.95	24.60	20.50	25.50	28.75	30.45
Saint Louis	22.35	32.70	29.95	30.20	32.40	36.20
Pacific	31.60	38.80	37.40	36.80	40.25	42.25
Sedalia	35.30	41.45	40.60	42.35	48.40	52.80
Kansas City	40.25	44.65	42.25	46.95	50.35	54.65
Topeka	52.75	53.00	50.00	52.60	56.75	58.85
Salina	58.20	66.05	62.80	58.20	61.80	66.30
Denver	65.00	72.50	68.20	62.25	65.50	74.25

24. EXTENDED OLD STYLE, Period 3-Point set.
Figure 1 cast 3½-Point or 6-Point set.
25. EXTENDED WOODWARD, Period 3-Point set.
Figure 1 cast 4-Point or 6-Point set.
26. LATIN ANTIQUE, Period 3-Point set.
Figure 1 cast 4-Point or 6-Point set.

27. Mixed Column, showing a variety of 4-Point
set Figures, illustrating how they agree in
justification. Special lines in tables may
thus be easily emphasized or differentiated
by the use of heavy-face Figures.
28. Mixed 5-Point sets. 29. Mixed 6-Point sets.

See also SPECIAL LIGHT-FACE and HEAVY-FACE TIME-TABLE FIGURES, on page 307.

8-POINT FIGURES, 4-POINT OR EN SET

30	31	32		33	34	35
12.00 am	7.05 pm	6.20 am	Le. ...Atchison... Ar.	5.05	11.25	5.20
12.23	7.25	6.40	.... Shannon.........	4.46	11.04	5.02
12.52	7.58	7.15	.... Everest	4.15	10.35	4.30
1.32	8.36	7.56	....Hiawatha.....	3.38	9.55	3.53
2.16	9.22	8.48	.... Falls City	3.06	9.13	3.21
2.55	9.50	9.26	.... Stella	2.26	8.35	2.40
3.30	10.25	10.00	.... Auburn	1.40	8.00	1.55
4.50	11.22	10.54	.... Wyoming	12.36	6.32	12.50
5.05	11.33	11.02	Ar...... Union ......Le.	12.23 am	6.20 pm	12.38 pm

30. ROMAN NO. 20, Period 3-Point set.
31. ITALIC NO. 20, Period 3-Point set.
32. OLD STYLE NO. 9, Period 3-Point set.
33. OLD STYLE ITALIC NO. 9, Period 3-Point.

34. KELMSCOTT, Period 2-Point set.
3-Point set Period cast to order if wanted.
35. SAINT JOHN, Period 2-Point set.
3-Point set Period cast to order if wanted.

8-POINT FIGURES, 5-POINT SET

	36	37	38	39	40	41	42
Pittsburg	$ 9.35	$ 8.25	$ 9.45	$10.75	$12.20	$13.35	$14.00
Cincinnati	10.25	9.60	10.10	14.60	16.45	17.50	18.80
Indianapolis	13.60	12.40	13.25	18.20	20.00	22.45	24.75
Saint Louis	21.45	18.75	19.95	24.25	26.80	28.00	30.10
Sedalia	28.75	25.80	27.50	30.95	32.25	35.20	36.80
Kansas City	36.20	30.20	33.20	39.15	41.25	44.75	45.20
Atchison	40.15	38.45	39.30	42.70	46.50	48.25	50.35
Lincoln	47.90	42.90	45.35	48.85	50.85	52.80	53.60
Omaha	52.00	50.10	51.75	54.40	55.75	56.60	57.25

36. ROMAN NO. 20, Wide Fig.
37. CONDENSED NO. 1.
38. CONDENSED NO. 2.

39. LATIN.
40. GOTHIC NO. 6.
41. GOTHIC ITALIC NO. 1.

42. Mixed Column, displaying
these Figures in combina-
tion under one another.
See note for column 27 above.

Periods of all these fonts are 3-Point set.

Do Not Order by these Index Numbers.

POINT-SET FIGURES

For Time-Tables, Tariffs, Etc.

8-POINT FIGURES, 6-POINT OR 3-4 EM SET

43	44	45	46	47	48	49	50
10.25	**25.30**	9.75	24.65	15.20	8.40	12.45	**23.50**
32.40	**15.65**	23.60	9.45	23.50	27.85	24.65	**6.30**
4.75	**39.00**	18.40	16.80	6.25	38.35	9.00	**35.00**
62.15	**8.25**	30.25	38.40	32.90	16.20	40.65	**21.35**
33.80	**41.85**	62.15	41.30	16.40	48.50	17.75	**46.50**
16.35	**26.10**	5.35	63.66	57.95	21.75	38.40	**8.95**
7.50	**52.75**	34.20	7.35	72.15	9.39	63.50	**36.75**
24.60	**20.60**	17.55	36.10	8.00	30.00	29.85	**15.25**
48.30	**5.50**	82.00	64.50	26.75	62.45	8.60	**52.10**

43. LIGHT-FACE No. 1, Period 3-Point set.
44. FULL-FACE No. 1, Period 3-Point set.
45. SKINNER, Period 2-Point set.
   3-Point set Period cast to order if wanted.
46. WOODWARD, Period 3-Point set.
   Figure 1 cast 3-Point or 6-Point set.
See column 58 for mixed display of 6-Point sets.

47. HALF-TITLE, Period 4-Point set.
   3-Point set Period cast to order if wanted.
48. GOTHIC No. 1, Period 3-Point set.
   Figure 1 cast 4-Point or 6-Point set.
49. ANTIQUE No. 1, Period 3-Point set.
50. EDWARDS, Period 3-Point set.
   Figure 1 cast 3½-Point or 6-Point set.

8-POINT FIGURES, VARIOUS POINT SETS

51	52	53	54	55	56	57	58
20.35	9.80	14.55	6.75	8.50	**23.60**	**26.30**	364.75
15.25	24.50	28.75	24.60	24.95	**10.40**	**34.95**	96.45
36.40	15.00	9.30	38.40	32.75	**45.25**	**25.45**	**240.30**
8.75	32.65	31.40	16.95	13.80	**6.35**	**48.69**	**793.50**
48.45	6.25	62.25	44.70	7.20	**20.15**	**32.50**	539.65
10.60	27.75	6.10	9.30	46.35	**57.20**	**57.95**	928.35
6.30	10.60	22.70	32.45	21.45	**5.80**	**72.60**	49.84
21.85	52.90	43.65	26.35	52.00	**31.45**	**64.25**	**326.70**
34.00	48.35	13.00	76.00	37.60	**62.00**	**9.30**	84.20

51. CONDENSED WOODWARD,
   Figures 3-Point set, Period 2-Point set.
52. CONDENSED GOTHIC No. 1,
   Figures 3½-Point set, Period 2-Point set.
53. TUDOR BLACK,
   Figures 5½-Point set, Period 3-Point set.
54. LATIN ANTIQUE,
   Figures 7-Point set, Period 3-Point set.
   Figure 1 cast 4-Point or 7-Point set.

55. EXTENDED OLD STYLE,
   Figures 8-Point set, Period 3-Point set.
   Figure 1 cast 5-Point or 8-Point set.
56. EXTENDED WOODWARD,
   Figures 8-Point set, Period 3-Point set.
   Figure 1 cast 4-Point or 8-Point set.
57. Mixed Column, displaying lines of Figures
   on 8-Point set under one another.
58. Mixed Column of Figures on 6-Point set.

Do Not Order by these Index Numbers.

5-POINT FIGURES

ROMAN No. 20 ....Figures 2½, Period 2
ITALIC No. 23..............Figures 3, Period 2
ROMAN No. 20, Wide....Figures 3½, Period 2
GOTHIC No. 6 ....Figures 3½, Period 2½
TITLE GOTHIC No. 72...Figures 4, Period 3
TITLE GOTHIC No. 52...Figures 5, Period 3

5½-POINT FIGURES

ITALIC No. 23..............Figures 3, Period 2
ROMAN No. 23, Wide....Figures 3½, Period 2
CONDENSED No. 1.......Figures 4, Period 3
FULL-FACE No. 1.........Figures 4, Period 3
HALF-TITLE................Figures 5, Period 3

7-POINT FIGURES

ROMAN No. 20.............Figures 3½, Period 2½
ROMAN No. 20, Wide....Figures 4½, Period 2½
ITALIC No. 20 ....Figures 3½, Period 2½
OLD STYLE No. 9........Figures 3½, Period 2½
OLD STYLE ITALIC No.9 Figures 3½, Period 2½
CONDENSED No. 1.......Figures 4, Period 3
CONDENSED No. 2.......Figures 4, Period 3
LATIN ....Figures 4, Period 3
CLARENDON ....Figures 4½, Period 2½
ANTIQUE No. 1...........Figures 5, Period 3
GOTHIC No. 1 ....Figures 5, Period 3
IONIC ....Figures 5, Period 3
WOODWARD ....Figures 5, Period 3
LATIN ANTIQUE.........Figures 6, Period 3

POINT-SET FIGURES

For Time-Tables, Tariffs, Etc.

JUSTIFIERS AND LOGOTYPES ON POINT SETS

Special Justifiers, or Spaces cast the same set as Figures or Periods, will be cast to order on any body and width desired, at same prices as ordinary spaces. But all our regular spaces, being cast on point sets, can be combined and used for the same purpose. Special-width Quads also furnished.

Logotypes for Time-Tables and other special work made to order. Prices on application. See page 321 for Logotypes for which we have matrices on hand.

9-POINT FIGURES, VARIOUS POINT SETS

ROMAN NO. 20	Figures 4½, Period 3	LATIN	Figures 5, Period 3
ITALIC NO. 20	Figures 4½, Period 3	FULL-FACE NO. 1	Figures 6, Period 3
OLD STYLE NO. 9	Figures 4½, Period 3	FULL-FACE NO. 2	Figures 7, Period 3½
OLD STYLE ITALIC NO. 9	Figures 4½, Period 3	3-Point set Period cast to order if wanted.	
CONDENSED NO. 1	Figures 5, Period 3	LATIN ANTIQUE	Figures 7, Period 3
CONDENSED NO. 2	Figures 5, Period 3	GOTHIC NO. 1	Figures 8, Period 3

10-POINT FIGURES, 4-POINT AND 5-POINT SETS

59	60	61	62	63	64	65	66	67	68
20.35	15.00	9.45	31.75	*2.60*	48.30	12.45	*23.65*	24.25	15.75
15.75	4.85	26.25	9.25	*12.85*	13.90	26.50	*8.85*	26.00	21.50
8.00	21.65	13.30	18.60	*30.00*	32.65	7.75	*36.20*	15.95	6.80
47.05	30.80	8.60	41.10	*26.30*	6.50	32.00	*4.90*	8.85	42.25
39.45	52.50	36.80	6.35	*52.25*	28.35	9.30	*13.35*	32.40	17.40
62.70	3.25	51.75	22.90	*7.50*	54.25	25.65	*47.25*	7.10	38.35
5.30	18.70	23.00	14.50	*15.80*	7.00	43.60	*21.00*	49.65	4.10

10-POINT FIGURES, 6-POINT AND 6½-POINT SETS

69	70	71	72	73	74	75	76
10.65	8.75	23.50	13.85	*6.40*	*10.25*	*20.15*	4.35
32.75	29.30	7.80	30.00	*33.90*	*38.00*	*8.75*	28.60
8.40	14.40	16.25	26.50	*14.65*	*6.90*	*35.00*	15.20
21.25	6.25	39.75	8.75	*27.30*	*21.75*	*16.25*	37.75
53.30	42.80	9.40	41.15	*9.25*	*46.35*	*42.80*	8.90
16.00	12.00	42.85	24.40	*41.00*	*62.20*	*7.35*	92.15
48.15	57.35	53.00	62.65	*76.20*	*15.85*	*59.45*	21.25

4-Point Set

59. COND. WOODWARD, Period 2-Point set.
60. COND. GOTHIC NO. 1, Period 2½-Point set.
61. COND. LATIN, Period 2½-Point set.
    For above three faces Periods will be cast either 2-Point or 2½-Point set to order.

5-Point Set

62. ROMAN NO. 20,
63. ITALIC NO. 20,
64. ROMAN NO. 23,
65. OLD STYLE NO. 9,
66. OLD STYLE ITALIC NO. 9,
67. FRENCH OLD STYLE,
68. KELMSCOTT, Period 2½-Point set.
    { Period 3½-Point set; will be cast on 3-Point set to order. }
    3-Point set Period cast to order if wanted.

6-Point Set

69. CONDENSED NO. 1,
70. CONDENSED NO. 2,
71. LATIN,
72. GOTHIC NO. 6,
    { Period 3½-Point set; will be cast to order on 3-Point set. }
73. GOTHIC ITALIC NO. 1, Period 3½-Point set.
    Figure 1 cast 4-Point or 6-Point set.
    3-Point set Period cast to order if wanted.
74. COSMOPOLITAN, Period 2½-Point set.
    3-Point set Period cast to order if wanted.

6½-Point Set

75. CALEDONIAN ITALIC, Period 3½-Point set.
    3-Point set Period cast to order if wanted.
76. ANTIQUE NO. 1, Period 3½-Point set.
    3-Point set Period cast to order if wanted.

Do Not Order by these Index Numbers.

POINT-SET FIGURES

For Tables, Rate Sheets, Calendars, Etc.

10-POINT FIGURES, 7-POINT, 8-POINT AND 9-POINT SETS

77	78	79	80	81	82	83
10.60	2.75	34.90	25.35	32.00	9.40	24.30
9.40	31.55	21.75	8.75	21.60	36.25	7.45
24.75	18.20	8.65	32.50	6.25	25.80	46.05
49.50	46.35	40.00	46.30	53.75	7.30	31.20
16.65	6.90	62.15	15.45	4.10	41.65	58.00
8.35	22.00	7.25	68.00	18.45	53.00	9.15
53.00	63.25	51.60	9.15	49.95	34.15	24.40

10-POINT FIGURES, 10-POINT SET, AND MIXED DISPLAY

84	85	86	87	88	89
27.50	24.00	62.70	25.80	46.30	35.95
52.85	6.25	19.45	62.25	34.85	26.90
9.65	52.70	23.75	91.30	53.75	28.60
10.00	31.80	48.90	83.90	82.60	6.70
38.15	49.65	37.60	27.85	26.50	44.35
5.25	12.85	53.35	54.65	45.80	8.24
46.80	68.50	10.30	37.45	27.25	45.50

7-Point Set

77. FULL-FACE NO. 1, ⎫ Period 3½-Point set;
78. TUDOR BLACK, ⎬ will be cast to order
79. WOODWARD,* ⎭ on 3-Point set.
　*Figure 1 cast 4-Point or 7-Point set.

8-Point Set

80. SKINNER, Period 2½-Point set.
　4-Point set Period cast to order if wanted.
81. LATIN ANTIQUE, Period 4½-Point set.
　4-Point set Period cast to order if wanted.
82. EDWARDS, Period 4-Point set,
　For above three faces the Figure 1 is cast
　either 5-Point or 8-Point set.

9-Point Set

83. GOTHIC NO. 1, Period 3½-Point set.
　4-Point set Period cast to order if wanted.
　Figure 1 cast 6-Point or 9-Point set.

10-Point Set

84. EXTENDED OLD STYLE, Period 3½-Point.
　Figure 1 cast 6-Point or 10-Point set.
　4-Point set Period cast to order if wanted.
85. EXTENDED WOODWARD, Period 3½-Point.
　4-Point set Period cast to order if wanted.
　Figure 1 cast 5-Point or 10-Point set.

Mixed Display

86. Display of various 6-Point set Figures used
　in combination in the same column, illus-
　trating the agreement in justification of
　the different styles. Special lines in tab-
　ular matter may therefore be emphasized
　by use of different Figures, without any
　trouble whatsoever.
87. Mixed display of 7-Point set Figures.
88. Mixed display of 8-Point set Figures.
89. Mixed display of 10-Point set Figures.

Do Not Order by these Index Numbers.

FIGURE 1—COMMA AND PERIOD—LEADERS

In a number of fonts the Figure 1 is ordinarily cast thinner than the other Figures, being easily justified by means of our spaces to occupy the same space as the other figures, if necessary. When a large quantity of Figures is ordered the 1, IF REQUESTED, will be cast the same width as the others. The Comma in all fonts is cast the same set or thickness as the Period.

Being cast on Standard Line, our Leaders can be used on tabular work in connection with every style of face and figure shown in these specimens. We make Standard Line Leaders in four styles:

No. 1, Round, 2 dots to em, from 5-Point to 18-Point, in en, em, 1½-em, 2-em and 3-em widths.
No. 2, Fine-dot, from 6-Point to 14-Point, in en, em, 1½-em, 2-em and 3-em widths.
No. 3, Round, 1 dot to em, from 6-Point to 12-Point, in em, 2-em and 3-em widths.
No. 4, Hyphen, from 5-Point to 12-Point, in en, em, 1½-em, 2-em and 3-em widths.
Leaders of special styles and widths made to order.

POINT-SET FIGURES

For Tables, Rate Sheets, Calendars, Etc.

12-POINT

CONDENSED WOODWARD, Figures 4½, Period 3

$12,345,678.90

CONDENSED GOTHIC NO. 1, Figures 5, Period 3

$12,345,678.90

CONDENSED LATIN   Figures 5, Period 3

$12,345,678.90

ROMAN NO. 20   Figures 6, Period 4

12,345,678.90

ITALIC NO. 20   Figures 6, Period 4

$12,345,678.90

OLD STYLE NO. 9   Figures 6, Period 4

$12,345,678.90

OLD STYLE ITALIC NO. 9   Figures 6, Period 4

$12,345,678.90

FRENCH OLD STYLE   Figures 6, Period 4

$12,345,678.90

FRENCH OLD STYLE ITAL., Figures 6, Period 4

$12,345,678.90

CONDENSED NO. 2   Figures 6, Period 4

$12,345,678.90

SCHWABACHER   Figures 6, Period 4

$12,345,678.90

KELMSCOTT   Figures 6, Period 3
Period cast 4-Point set to order.

$12,345,678.90

SAINT JOHN   Figures 6, Period 3
Period cast 4-Point set to order.
Figure 1 cast 4-Point or 6-Point set.

$12,345,678.90

CONDENSED NO. 1   Figures 7, Period 4

$12,345,678.90

LATIN   Figures 7, Period 4

$12,345,678.90

12-POINT

GOTHIC NO. 6   Figures 7, Period 3

$12,345,678.90

CALEDONIAN ITALIC   Figures 7, Period 3
Period cast 4-Point set to order.

$12,345,678.90

GOTHIC ITALIC NO. 1   Figures 7, Period 4
Figure 1 cast 4-Point or 6-Point set.

$12,345,678.90

COSMOPOLITAN   Figures 7, Period 3
Period cast 4-Point set to order.

$12,345,678.90

FULL-FACE NO. 1   Figures 8, Period 4

$12,345,678.90

IONIC   Figures 8, Period 4

$12,345,678.90

TUDOR BLACK   Figures 8, Period 4

$12,345,678.90

WOODWARD   Figures 8, Period 4
Figure 1 cast 5-Point or 8-Point set.

$12,345,678.90

ANTIQUE NO. 1   Figures 9, Period 4
Figure 1 cast 6-Point or 9-Point set.

$12,345,678.90

EDWARDS   Figures 9, Period 4
Figure 1 cast 5-Point or 9-Point set.

$12,345,678.90

SKINNER   Figures 9, Period 3
Period cast 4-Point set to order.
Figure 1 cast 5-Point or 9-Point set.

$12,345,678.90

GOTHIC NO. 1   Figures 10, Period 4
Figure 1 cast 6-Point or 10-Point set.

$12,345,678.90

EXTENDED OLD STYLE   Figures 11, Period 4
Figure 1 cast 7-Point or 11-Point set.

$12,345,678.90

EXTENDED WOODWARD   Figures 12, Period 4
Figure 1 cast 6-Point or 12-Point set.

$1,234,567.89

POINT-SET FIGURES

For Tables, Rate Sheets, Calendars, Etc.

14-POINT

CONDENSED WOODWARD. Figures 6, Period 3
Period cast 4-Point set to order.

$12,345,678.90

CONDENSED GOTHIC NO. 1, Figures 6, Period 4

$12,345,678.90

KELMSCOTT   Figures 7, Period 3
Period cast 4-Point set to order.

$12,345,678,90

SAINT JOHN   Figures 7, Period 4
Figure 1 cast 5-Point or 7-Point set.

$12,345,678.90

CONDENSED GOTHIC NO. 4, Figures 7, Period 4

$12,345,678.90

COSMOPOLITAN   Figures 8, Period 3
Period cast 4-Point set to order.

$12,345,678.90

CONDENSED NO. 2   Figures 8, Period 4

$12,345,678.90

GOTHIC ITALIC NO. 1   Figures 8, Period 5
Period cast 4-Point set to order.
Figure 1 cast 6-Point or 8-Point set.

$12,345,678.90

GOTHIC NO. 6   Figures 9, Period 4
Period cast 5-Point set to order.

$12,345,678.90

WOODWARD   Figures 9, Period 5
Figure 1 cast 5-Point or 9-Point set.

$12,345,678.90

SKINNER   Figures 11, Period 3
Period cast 5-Point set to order.
Figure 1 cast 7-Point or 11-Point set.

$12,345,678.90

EDWARDS   Figures 11, Period 5
Figure 1 cast 7-Point or 11-Point set.

$12,345,678.90

GOTHIC NO. 1   Figures 12, Period 5
Figure 1 cast 9-Point or 12-Point set.

$1,234,567.89

14-POINT

EXTENDED WOODWARD   Figures 11, Period 5
Figure 1 cast 8-Point or 14-Point set.

$123,456.78

18-POINT

CONDENSED LATIN   Figures 6, Period 4

12,345,678.90

CONDENSED GOTHIC NO. 1, Figures 7, Period 4

12,345,678.90

CONDENSED WOODWARD   Figures 7, Period 4

12,345,678.90

CONDENSED NO. 2   Figures 8, Period 4

12,345,678.90

KELMSCOTT   Figures 9, Period 4
Period cast 5-Point set to order.

12,345,678.90

SAINT JOHN   Figures 9, Period 5
Figure 1 cast 6-Point or 9-Point set.

12,345,678.90

LATIN   Figures 10, Period 6

12,345,678.90

GOTHIC ITALIC NO. 1   Figures 11, Period 6
Figure 1 cast 7-Point or 11-Point set.

12,345,678.90

IONIC   Figures 12, Period 6

12345,678.90

DORIC   Figures 12, Period 6

1234567890

WOODWARD   Figures 12, Period 7
Period cast 6-Point set to order.
Figure 1 cast 8-Point or 12-Point set.

12345,678.90

POINT-SET FIGURES

For Tables, Rate Sheets, Calendars, Etc.

18-POINT

TUDOR BLACK   Figures 12, Period 4
Period cast 6-Point set to order.
Figure 1 cast 8-Point or 12-Point set.

12,345,678.90

COSMOPOLITAN   Figures 12, Period 4
Period cast 6-Point set to order.
Figure 1 cast 8-Point or 12-Point set.

12,345,678.90

LATIN ANTIQUE   Figures 13, Period 6
Figure 1 cast 7-Point or 13-Point set.

1,234,567.89

ANTIQUE No. 1   Figures 13, Period 5
Period cast 6-Point set to order.
Figure 1 cast 9-Point or 13-Point set.

1,234,567.89

SKINNER   Figures 14, Period 4
Period cast 6-Point set to order.
Figure 1 cast 8-Point or 14-Point set.

1,234,567.89

EDWARDS   Figures 14, Period 6
Figure 1 cast 8-Point or 12-Point set.

1234,567.89

GOTHIC No. 1   Figures 14, Period 6
Figure 1 cast 9-Point or 14-Point set.

123,456.78

EXTENDED OLD STYLE   Figures 15, Period 6
Figure 1 cast 9-Point or 15-Point set.

123,456.78

EXTENDED WOODWARD   Figures 18, Period 6
Figure 1 cast 10-Point or 18-Point set.

12,345.67

24-POINT

CONDENSED GOTHIC No. 1, Figures 8, Period 4

12,345,678.90

24-POINT

CONDENSED LATIN   Figures 9, Period 5
Figure 1 cast 8-Point or 9-Point set.

12,345,678.90

CONDENSED WOODWARD   Figures 9, Period 4
Period cast 5-Point set to order.

12,345,678.90

LATIN   Figures 12, Period 6

12345,678.90

SAINT JOHN   Figures 12, Period 6
Figure 1 cast 8-Point or 12-Point set.

12345,678.90

KELMSCOTT   Figures 12, Period 5
Period cast 6-Point set to order.

12345,678.90

CONDENSED No. 2   Figures 12, Period 5
Period cast 6-Point set to order.

12345,678.90

COSMOPOLITAN   Figures 13, Period 6
Figure 1 cast 9-Point or 13-Point set.

1,234,567.89

GOTHIC ITALIC No. 1   Figures 14, Period 6
Figure 1 cast 10-Point or 14-Point set.

1234567.89

TUDOR BLACK   Figures 14, Period 5
Period cast 6-Point set to order.
Figure 1 cast 10-Point or 14-Point set.

1234,567.89

FRENCH OLD STYLE   Figures 14, Period 6

123,456.78

POINT-SET FIGURES

For Tables, Rate Sheets, Calendars, Etc.

24-POINT

WOODWARD   Figures 15, Period 8
Figure 1 cast 11-Point or 15-Point set.

123,456.78

LATIN ANTIQUE   Figures 16, Period 7
Figure 1 cast 8-Point or 16-Point set.

123456.78

GOTHIC No. 1   Figures 18, Period 6
Figure 1 cast 10-Point or 18 Point set.

12,345.67

GOTHIC No. 6   Figures 18, Period 6
Figure 1 cast 10-Point or 18-Point set.

12,345.67

SKINNER   Figures 18, Period 4
Period cast 6-Point set to order.
Figure 1 cast 12-Point or 18-Point set.

12,345.76

EDWARDS   Figures 18, Period 8
Figure 1 cast 12-Point or 18-Point set.

12345.67

EXTENDED OLD STYLE   Figures 20, Period 6
Figure 1 cast 12-Point or 20-Point set.

1,234.56

EXTENDED WOODWARD   Figures 22, Period 8
Figure 1 cast 12-Point or 22-Point set.

1234.56

30-POINT

CONDENSED GOTHIC No. 1 Figs. 10, Period 6

12,345,678.90

30-POINT

CONDENSED WOODWARD, Figures 11, Period 5
Figure 1 cast 7-Point or 11-Point set.

12,345,678.90

KELMSCOTT   Figures 15, Period 6

123,456.78

LATIN   Figures 16, Period 7
Period cast 8-Point set to order.

12,345.67

WOODWARD   Figures 16, Period 8
Figure 1 cast 12-Point or 16-Point set.

12,345.67

TUDOR BLACK   Figures 16, Period 8
Figure 1 cast 12-Point or 16 Point set.

12,345.67

COSMOPOLITAN   Figures 18, Period 6
Figure 1 cast 14-Point or 18-Point set.

12345.67

SKINNER   Figures 22, Period 5
Figure 1 cast 14-Point or 22-Point set.

1234.56

EDWARDS   Figures 22, Period 9
Figure 1 cast 12-Point or 22 Point set.

1234.56

**All Figures and Points of faces on
larger bodies are also on Point sets.**

FRACTION SPECIMENS

Fractions are not supplied with Roman fonts unless especially ordered. In ordering, be particular to state the No. of the Fractions wanted, as well as the body and quantity. Fractions are sold at the same prices as Roman; see page 11. They are also put up in 1-pound fonts, at following prices:

6-POINT No. 1 — En set 64c.	6-POINT No. 3 — Em set 64c.
7-POINT No. 1 — En set 56c.	7-POINT No. 3 — Em set 56c.
8-POINT No. 1 — En set 53c.	8-POINT No. 3 — Em set 53c.
9-POINT No. 1 — En set 50c.	9-POINT No. 3 — Em set 50c.
10-POINT No. 1 — En set 48c.	10-POINT No. 3 — Em set 48c.
11-POINT No. 1 — En set 46c.	11-POINT No. 3 — Em set 46c.
12-POINT No. 1 — En set 45c.	6-POINT No. 9 — Em set 64c.
5-POINT No. 2 — Em set $1.20	8-POINT No. 9 — Em set 53c.
5½-POINT No. 2 — Em set 74c.	9-POINT No. 9 — Em set 50c.
6-POINT No. 2 — Em set 64c.	10-POINT No. 9 — Em set 48c.
7-POINT No. 2 — Em set 56c.	11-POINT No. 9 — Em set 46c.
8-POINT No. 2 — Em set 53c.	12-POINT No. 9 — Em set 45c.
9-POINT No. 2 — Em set 50c.	

MISCELLANEOUS AUXILIARIES

PIECE FRACTIONS

Put up in 1-pound fonts, at the prices given.

8-POINT $3.60

1234567890 __ 1234567890

$\frac{270}{353}$ $\frac{380}{027}$ $\frac{410}{460}$

10-POINT $2.80

1234567890 __ 1234567890

$\frac{132}{512}$ $\frac{548}{873}$ $\frac{709}{940}$

12-POINT $2.00

1234567890 __ 1234567890

$\frac{168}{457}$ $\frac{257}{372}$ $\frac{430}{916}$

11-POINT also cast by us; per font, $2.40

SUPERIOR AND INFERIOR LETTERS AND FIGURES

The following Superior and Inferior Letters and Figures are put up in 1-pound fonts, at the prices given. Letters and Figures are in separate fonts.

6-POINT $2.00

mgd^{abcdefghijklmnopqrst} mgd^{1234567890}

mdg$_{abcdefghijklmnopqrst}$ mdg$_{1234567890}$

7-POINT $1.80

mgd^{abcdefghijklmnopqru} mgd^{1234567890}

mdg$_{abcdefghijklmnopqru}$ mdg$_{1234567890}$

8-POINT $1.60

mgd^{abcdefghijklmnopqrs} mgd^{1234567890}

mdg$_{abcdefghijklmnopqrs}$ mdg$_{1234567890}$

9-POINT $1.44

mgd^{abcdefghijklmnop} mgd^{1234567890}

mdg$_{abcdefghijklmnop}$ mdg$_{1234567890}$

10-POINT $1.30

mgd^{abcdefghijklmnw} mgd^{1234567890}

mdg$_{abcdefghijklmnw}$ mdg$_{1234567890}$

11-POINT $1.22

mgd^{abcdefghijklmnt} mgd^{1234567890}

mdg$_{abcdefghijklmnt}$ mdg$_{1234567890}$

12-POINT $1.16

mgd^{abcdefghijkn} mgd^{1234567}

mdg$_{abcdefghijkn}$ mdg$_{1234567}$

ARITHMETICAL SIGNS

Prices of fonts.

5½-POINT $2.00

$+ \ - \ \times \ \div \ =$

6-POINT $1.50

$+ \ - \ \times \ \div \ =$

7-POINT $1.50

$+ \ - \ \times \ \div \ =$

8-POINT $1.25

$+ \ - \ \times \ \div \ =$

9-POINT $1.20

$+ \ - \ \times \ \div \ =$

10-POINT $1.00

$+ \ - \ \times \ \div \ =$

11-POINT $1.00

$+ \ - \ \times \ \div \ =$

12-POINT $1.00

$+ \ - \ \times \ \div \ =$

SUPERIOR AND INFERIOR FRACTIONS

Superior and Inferior Fractions are put up in 1-pound fonts, at the prices given.

6-POINT $2.00

$3^{21}/_{43}$ $5^5/_9$ $^{1234567890}/_{1234567890}$ $6^3/_5$ $7^{20}/_{61}$

7-POINT $1.80

$13^{562}/_{9250}$ $^{1234567890}/_{1234567890}$ $42^{730}/_{8325}$

8-POINT $1.60

$78^{748}/_{925}$ $^{1234567890}/_{1234567890}$ $90^{232}/_{683}$

9-POINT $1.44

$54^{36}/_{9265}$ $^{1234567890}/_{1234567890}$ $63^{37}/_{4802}$

10-POINT $1.30

$21^{394}/_{573}$ $^{1234567890}/_{1234567890}$ $90^{247}/_{366}$

11-POINT $1.22

$75^{54}/_{306}$ $^{1234567890}/_{1234567890}$ $86^{34}/_{679}$

12-POINT $1.16

$48/_{75}$ $^{1234567890}/_{1234567890}$ $23/_{26}$

MISCELLANEOUS AUXILIARIES

COMMERCIAL MARKS
Prices per pound.

5-POINT $1.20

℔ @ ℔ ⅌ ⅌ ¢

5½-POINT   74c.

℔ @ ℔ ⅌ ⅌ ¢

6-POINT   64c.

℔ @ ℔ ⅌ ⅌ ¢

7-POINT   56c.

℔ @ ℔ % ℀ ¢

8-POINT   53c.

℔ @ ℔ % ℀ ¢

9-POINT   50c.

℔ @ ℔ % ℀ ¢

10-POINT   48c.

℔ @ ℔ % ℀ ¢

11-POINT   46c.

℔ @ ℔ % ℀ ¢

12-POINT   45c.

℔ @ ℔ % ℀ ¢

ALGEBRAIC AND GEOMETRICAL SIGNS
Prices of fonts.

5½-POINT $2.00

$+ - \times \div = \div : :: \pm \square \square \triangle <$
$\sqrt{} \sqrt[3]{} \angle$

6-POINT $1.50

$+ - \times \div = \div : :: \pm \square \square \triangle$
$< \sqrt{} \sqrt[3]{} \angle$

7-POINT $1.50

$+ - \times \div = \div : :: \pm \square \square$
$\triangle < \sqrt{} \sqrt[3]{} \angle$

8-POINT $1.25

$+ - \times \div = \div : :: \pm \square \triangle <$
$\sqrt{} \sqrt[3]{} \angle \frown$

9-POINT $1.20

$+ - \times \div = \div : :: \pm \square \triangle$
$< \sqrt{} \sqrt[3]{} \angle \frown$

ALGEBRAIC AND GEOMETRICAL SIGNS
Prices per pound.

10-POINT $1.00

$+ - \times \div = \div : :: \pm \bigcirc \square$
$\triangle < \sqrt{} \sqrt[3]{} \angle \frown$

11-POINT $1.00

$+ - \times \div = \div : :: \pm \bigcirc \square$
$\triangle < \sqrt{} \sqrt[3]{} \angle \frown$

12-POINT $1.00

$+ - \times \div = \div : :: \pm \bigcirc$
$\square \triangle < \sqrt{} \sqrt[3]{} \angle \frown$

MEDICAL SIGNS
Prices of fonts.

6-POINT   50c.

℞   ℈   ℥   ℨ
6505   6506   6507   6508

7-POINT   50c.

℞   ℈   ℥   ℨ
7505   7506   7507   7508

8-POINT   50c.

℞   ℈   ℥   ℨ
8504   8505   8506   8507

9-POINT   50c.

℞   ℈   ℥   ℨ
9504   9505   9506   9507

10-POINT   50c.

℞   ℈   ℥   ℨ
10504   10505   10506   10507

11-POINT   50c.

℞   ℈   ℥   ℨ
11504   11505   11506   11507

12-POINT   50c.

℞   ℈   ℥   ℨ
12504   12505   12506   12507

Following Recipe Marks are sold singly.

℞   ℞   ℞   ℞
24506   18507   14501   12504
10c.   10c.   6c.   5c.

MISCELLANEOUS AUXILIARIES

NEW QUOTATION MARKS AND DASHES

IMPROVED DASHES AND
DOUBLE QUOTATION MARKS

For explanation of the special merits of these new Dashes and Quotation Marks see page 16.

The Improved Dashes are put up separately in 1-pound fonts, at the prices given.

The Double Quotation Marks are also put up in 1-pound fonts, containing an equal number of each, at the prices given.

5-POINT $1.20
— " " "Hunchback of Notre Dame"—Hugo.

5½-POINT 74c.
— " " "David Copperfield"—Dickens.

6-POINT 64c.
— " " "Huckleberry Finn"—Mark Twain.

7-POINT 56c.
— " " "King Lear"—Shakspeare.

8-POINT 53c.
— " " "Daniel Deronda"—Eliot.

9-POINT 50c.
— " " "Wandering Jew"—Sue.

10-POINT 48c.
— " " "Ben Hur"—Wallace.

11-POINT 46c.
— " " "Ivanhoe"—Scott.

12-POINT 45c.
— " " "Moths"—Ouida.

HEAVY QUOTATION MARKS

GUILLEMETS, OR FRENCH QUOTATION MARKS

The following style of Guillemets, for use with heavy job faces, will be furnished for all bodies. Sold at second-class prices.

«New Model» «Record»

«Devil»

«Bind» «Mail»

HEAVY QUOTATION MARKS

NEW IDEAL QUOTATION MARKS

The following is another style of Quotation Mark, intended to meet the ideas of many printers who do not like the customary method of quoting used in English typography and yet are not satisfied to adopt the French marks. These may be likened to the German marks, yet will be found to be quite different. They will be made for all bodies in addition to the following. Sold at second-class prices.

₂Fast Express₂ ₂Democrat₂

₂Milch₂

₂Saint₂ ₂David₂

Either of the above styles of Quotation Marks will be cut and cast to order for Roman fonts.

ACCENTS

NOTE: The German, French, Spanish and Swedish Accents are supplied to order in the lower case of all our Roman faces. For the majority of our Job faces we can furnish Spanish lower case Accents.

Accents and all styles of Marked Letters will be cut to order, the cost of making each matrix being from $2.00 to $5.00; type cast from same at prices given on page 11.

We furnish no quantity for less than 25c. net.

PIECE ACCENTS

For use with Job Faces. Prices per font.

3-POINT 50c.

4-POINT 50c.

6-POINT 50c.

10-POINT 50c.

14-POINT 50c.

MISCELLANEOUS AUXILIARIES

MISCELLANEOUS SIGNS AND LOGOTYPES

Prices per pound.

5-POINT	$2.60

o / " " / ẟ AM PM AR. LE

5½-POINT	$2.40

o / " " / ẟ doz

6-POINT	$2.00

o / " " / doz AM PM AR. LE. AM PM
via am pm AM PM Ar. Le. Ar. Le. am pm
do do No. Ar. Lv. A.M. P.M. ẟ §

7-POINT	$1.80

o / " " / AM PM

8-POINT	$1.60

o / " / / ◆ * §

9-POINT	$1.44

o / " /

10-POINT	$1.30

o / \ /

11-POINT	$1.22

o / \ /

12-POINT	$1.16

o \ / /

Other Miscellaneous Signs will be cut to order, the cost of making each matrix being from $2.00 to $5.00; type cast from same at prices given on page 11.
We furnish no quantity for less than 25c.

SPECIAL LOGOTYPES

LOGOTYPE SPECIMENS

The Evening Tribune
minimum rising snow storm
change has moderate Sunday
Alabama Arizona Erie Upper
barometer districts Nebraska

Special Logotypes of every description will be promptly cut and cast to order. Prices of such work sent on application.

IMPRINT LOGOTYPES

ELECTROTYPED IMPRINTS, each, 25 to 50c.

No. 52
CARSON-HARPER PRINT.

No. 53
WALLE & CO., NEW ORLEANS.

No. 54
GREAT WESTERN PTG. CO., ST. LOUIS.

No. 71
BUXTON & SKINNER PRINT.

No. 72
F. J. SCHUSTER, PRINTER.

No. 73
KEYSTONE PRESS, WELLSTON, O.

No. 74
R. P. STUDLEY & CO., PRINTERS, ST. LOUIS.

TYPE-METAL IMPRINTS

Imprints will be cast in Type-Metal, in any length under 72 points, and in any quantity over 50 of each, at following prices:

100 of any one style, $8.00 net.
50 of any one style, 5.00 net.

If Imprints are wanted longer than 72 points, they will have to be made in two or more sections, and $3.00 per 100 will be charged for each extra section.

JOB FACE CENT MARKS

¢   ¢   ¢   ¢
12510 14502 18510 24513

	Body	Each	Per pound
No. 12510	12-Point	5c	$1.16
No. 14502	14-Point	5c	1.12
No. 18510	18-Point	6c	1.00
No. 24513	24-Point	10c	.90

ELECTION TICKET SIGNS

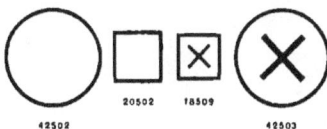

○ □ ⊠ ⊗
42502   20502 18509   42503

	Body	Each
No. 18509	18-Point	6c.
No. 20502	20-Point	6c.
No. 42502	42-Point	15c.
No. 42503	42-Point	15c.

IMPROVED METAL BRACES

IMPROVED METAL BRACES—ALL CAST ON 6-POINT BODY

Cast in following lengths from 32-Point to 72-Point—Price, 5c. each.
Put up in fonts of 2 of each length; including sectional pieces; per font, $1.00.

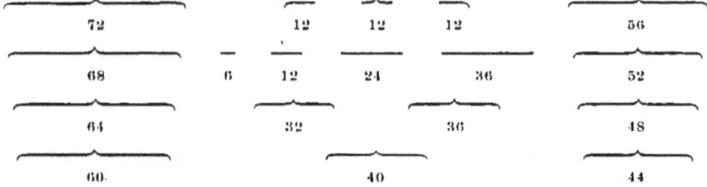

72	12	12	12	56
68	6	12	24	36	52
64	32		36		48
60		40		44

◆◆◆

SPACE RULES

Metal, cast on 2-Point set, on all bodies from 6-Point to 24-Point; per pound, $1.60.
Brass Space Rule is supplied on any body and any length.
Brass, 2-Point.........per pound, $2.00 to $2.50
Brass, 1½-Point ......per pound, 2.75 to 3.25
Brass, 1-Point........ per pound, 3.50 to 4.00

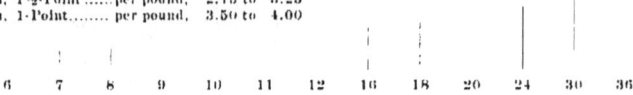

6   7   8   9   10   11   12   16   18   20   24   30   36

◆◆◆

MISCELLANEOUS CUTS

Cast in Type Metal. With a Few Exceptions all our Cuts are Multiples of 6-Point in Body and Width

12504—5c.   12508—5c.   12509—5c.   14501—6c.

18506—10c.   18507—10c.   18508—10c.   18511—10c.   18512—10c.   18513—10c.   18514—10c.

20501—10c.   24504—15c.   24505—10c.   24506—12c.   24507—10c.   24508—15c.

24509—10c.   24510—15c.   24511—15c.   24512—10c.   24514—10c.   24515—20c.

24517—15c.   30504—15c.   30505—15c.   30506—25c.   30507—15c.

MISCELLANEOUS CUTS

Cast in Type Metal

36504 — 20c.   36505 — 15c.   36506 — 15c.   36507 — 10c.   36508 — 15c.   36509 — 15c

36510 — 20c.   36511 — 20c.   36512 — 25c.   36513 — 15c.   36515 — 15c.

36517 — 20c.   36518 — 25c.   36519 — 25c.   36520 — 35c.

36521 — 40c.   36522 — 25c.   36523 — 25c.   30508 — 20c

42501 — 15c.   42504 — 35c.   42505 — 25c.   42506 — 35c.

42507 — 35c.   42508 — 25c.   42509 — 25c.   42510 — 25c.   48508 — 25c.

48504 — 20c.   48505 — 20c.   48506 — 30c.   48507 — 35c.

54501 — 20c.   54502 — 25c.   54503 — 25c.   54504 — 25c.   54509 — 25c.

MISCELLANEOUS CUTS

Cast In Type Metal

60501 — 40c.   72504 — 25c.   72505 — 40c.   72506 — 40c.

72507 — 40c.   72508 — 50c.   72509 — 40c.

72510 — 50c.   72511 — 50c.   72512 — 50c.

 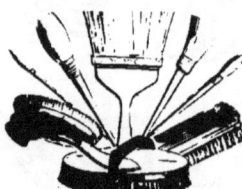 

72513 — 50c.   72514 — 50c.   72515 — 50c.

72516 — 50c.   72517 — 50c.   72518 — 50c.

MISCELLANEOUS CUTS

Cast in Type Metal

72519 — 50c.

72520 — 50c.

72521 — 50c.

72522 — 50c.

72523 — 50c.

72524 — 50c.

72525 — 50c.

72526 — 50c.

72527 — 50c.

72528 — 50c.

72529 — 50c.

72530 — 50c.

78501 — 50c.

90501 — 50c.

90502 — 50c.

96501 — 50c.

BLACK FISTS AND STARS

BLACK FISTS are put up in fonts containing two of each size; per font, $1.75

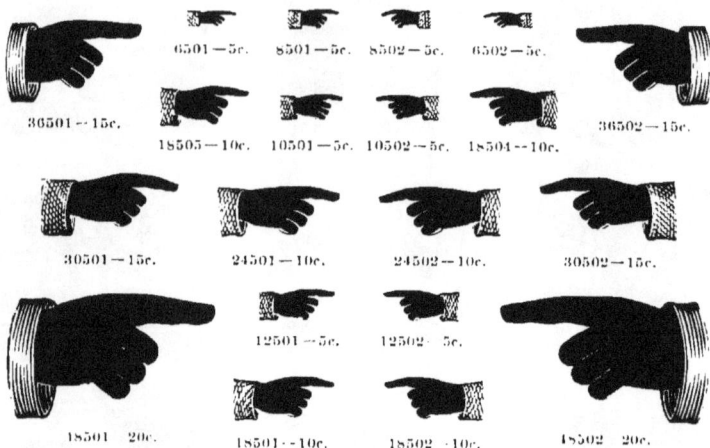

6501—5c.   8501—5c.   8502—5c.   6502—5c.

36501—15c.   36502—15c.

18505—10c.   10501—5c.   10502—5c.   18504—10c.

30501—15c.   24501—10c.   24502—10c.   30502—15c.

12501—5c.   12502—5c.

18501—20c.   18501—10c.   18502—10c.   18502—20c.

Single BLACK FISTS are sold separately at the prices given under each.

BLACK STARS are put up in fonts containing four of each—excepting Nos. 48503, 60503 and 72503, which are sold separately. Per font, $1.00.

6503—4c.   6504—4c.   8503—4c.   9503—4c.   10503—5c.   12503—5c.

36503—10c.   30503—8c.   24503—8c.   18503—5c.

48503—15c.

72503—25c.   60503—20c.

Single BLACK STARS are sold separately at the prices given under each.

ELECTROTYPED DATE-LINES

24-POINT COMMERCIAL SCRIPT DATE-LINE — 75c.

St. Louis, Mo., ...*189*

18-POINT COMMERCIAL SCRIPT DATE-LINE — 75c.

Indianapolis, Ind., ...*189*

24-POINT INVITATION SCRIPT DATE-LINE — 75c.

Springfield, Ill., ...*189*

18-POINT INVITATION SCRIPT DATE-LINE — 75c.

Fort Scott, Kans, ...*189*

12-POINT INVITATION SCRIPT DATE-LINE — 75c.

Kirkwood, Mo., ...*189*

24-POINT STATIONER SCRIPT DATE-LINE — 75c.

Dubuque, Iowa, ...*189*

24-POINT COSMOPOLITAN DATE-LINE — 75c.

Keokuk, Iowa, ...*189*

18-POINT COSMOPOLITAN DATE-LINE — 75c.

Galveston, Tex., ...*189*

14-POINT COSMOPOLITAN DATE-LINE — 75c.

Little Rock, Ark., ...*189*

12-POINT COSMOPOLITAN DATE-LINE — 75c.

Kansas City, Mo., ...*189*

N. B. — In ordering Date-Lines be particular to state what length you desire the dotted rule line to be. Unless otherwise ordered, we will make it 9 ems 12-Point. Dotted line will be added after the figures 189... only when so ordered, and the length of this rule stated.

SPECIMENS OF FACES FOR HEADLINES

MUCH ASTONISHED!

A Leading Printer Whose Ledger Showed
He Was Actually Making Money

ALMOST BEYOND BELIEF!

Further Investigation Develops the Cause of
His Establishment Giving Large Profits

REMARKABLE DISCOVERY MADE

Matter of Great Importance to the
Printing Trades, Which Have
Had a Hard Row to Hoe

WIDESPREAD SATISFACTION PROBABLE

Strange History of a Printer Who Bought
an Outfit of Standard Line Type—What
He Thinks of Its Money-Making Possibilities—Advice to Buyers of Material.

LEFT IN THE REAR

Old-Time Founders Unable to Catch the
Step of the Procession's Leaders

FIN DE SIECLE TYPES CAST

Progressive Young House Setting a Pace Hard for
the Decrepit Conservatives to Understand
and Much Harder for Them to Follow

QUICK COMPOSITORS LIKED

Our Point-Set Romans Help Them Greatly
in the Production of Lengthy Strings

LIST OF FACES FOR HEADS

Names and Sizes of the Display Letter in
the Headings Shown on this Page

First Column—24, 18, 14, 12 and 10-Point Conid-mard
Woodward, and 8 and 6-Point Woodward.

Second Column—1, 12, 10, 8 and 6-Point Conid-mard
No. 1.
2, 8 and 6-Point Gothic No. 1.
3, 8 and 6-Point Woodward.

Third Column—24, 18, 14, 12 and 10-Point Conid-mard
Gothic No. 1, and 8 and 6-Point Gothic No. 1.
For prices of regular fonts see series specimen pages.
These faces will also be furnished in fonts of 25 pounds
and over, at poster font prices.

CENTURY OF PROGRESS

Greatest Advance in the Art of Making Type
Brought About in Its Closing Years

NEW METHOD IN LINING TYPE

Original Inventions and Innovations Which Enlarge
the Money-Making Features of Typography

MOST USEFUL FACES CAST BY US

Favorite Styles of Newspaper, Book
and Jobbing Series Now Made
on Systematic Principles

STANDARD LINE AND UNIT SET POPULAR

Printers Everywhere Recognize the Merits
and Advantages Possessed by the Greatly
Improved Products Manufactured by the
Inland Type Foundry, of Saint Louis.

St. Louis, Mo., U. S. A.

INLAND TYPE FOUNDRY

SPECIMENS OF FACES FOR HEADLINES

PRINTERS CAUTIONED!

Spurious Pirating of Standard Line System is at Present Going On

IMITATIONS MOST WORTHLESS

Inability of the Copyists to Comprehend Our System—Futile Efforts Made to Devise Passable Substitutes for the Best

WELCOME NEWS TO PRINTERS

The Number and Variety of the Inland Type Foundry Products is Ever Increasing

THE STRIFE FOR PROFITS

Printers Aided In the Contest by Use of Standard Line Unit Set Type

STRONG PROOF OF ITS POSSIBILITY

Confounds the Opposers of Improvement in the Casting of Type as to Lining and Width

FIRST COLUMN—1. 12, 10, 8 and 6-Point Latin
2. 8 and 6-Point Condensed No. 2.
3. 8 and 6-Point Half-Title. 4. 6-Point Clarendon.
SECOND COLUMN—24, 18, 14, 12, 10, 8 and 6-Point Condensed No. 2.

FOUND RELIEF

A Prominent Printer on the Verge of Closing Shop

BUSINESS OUTLOOK

Most Discouraging Prospects of Bankruptcy Threaten Him

PROSPERITY VISIBLE

His Attention is Directed to a Medium by Which He Can Prevent Large Losses in His Type Setting

MAKE RIGHT BEGINNING

By Replacing Worthless Out-of-Date Material with Our Standard Line and Unit Set Types He Will Quickly be Cheered by Big Figures on the Profit Side of His Ledger Accounts.

QUICK GROWTH OF A BEGINNER

Rapid Progress of a New Type Foundry Causing Remark

ATTENTION TO PRINTERS' REQUIREMENTS

Why the Inland Type Foundry Made Such a Great Step to the Front a Problem Easily Explained

NINETEENTH CENTURY TYPES

Standard System for Coming Generations of Printers Has Already Appeared

MODERNIZED AIDS PROVIDED

Artistic Printers Helped Greatly in Composing Fine Work by Perfected Lining Faces

THE VALUE OF SYSTEM NOTED

Undeniable Advantages of Standard Line Type on Point Bodies and Unit Sets

THIRD COLUMN—1. 12 and 10-Point Condensed Latin, and 8 and 6-Point Latin Antique.
2. 8-Point Condensed No. 1, and 6-Point Full-Face No. 1.
3. 8 and 6-Point Gothic No. 6. 4. 6-Point Ionic.

ST. LOUIS, MO., U. S. A.

INLAND TYPE FOUNDRY

STANDARD LENGTH, 13 EMS 12-POINT. EACH, 50c. ORDER BY NUMBER. OTHER FACES MAY BE SELECTED FROM OUR SPECIMEN BOOK.

No. 51
DAILY LEADER

No. 52
MORTON EXPRESS

No. 53
Morning Courier

No. 54
BOSTON LEADER

No. 55
Norwich Democrat

No. 56
JOHNSON DISPATCH

No. 57
Mound City Republican

No. 58
BLUEFIELD HERALD BUCKNER SUN

No. 59
BLACKFORD TELEGRAPH

No. 60
Moniteau Advertiser

No. 61
TRENCHFIELD ADVOCATE

No. 62
PINEWOOD MORNING GAZETTE

No. 63
Sedalia Evening Chronicle

No. 71
Sunday Morning Tribune

No. 64
MORRISON EAGLE

No. 65
Garfield Patriot

No. 66
Morning Review

No. 67
Grand Pass Advance

No. 68
Atchison Blade

No. 69
Weekly Transcript

No. 70

St. Louis, Mo., U. S. A.

330

Prices of Headings given on application   No. 1. 48-Point Newspaper Title, Caps only   Special styles will be engraved to order

FOSTER TIMES

No. 2. 60-Point Cosmopolitan, Caps and Lower Case

Rochester Beacon

No. 3. 48-Point Cosmopolitan, Caps and Lower Case

Kirkwood Intelligencer

No. 4. 48-Point Tudor Black, Caps and Lower Case

Little Rock Democrat

DISTON NEWS

Urbana Express

HELMICK LEADER

Belleville Champion

St. Louis, Mo., U. S. A.

No. 9. 60-Point Condensed Woodward, Caps

BOSTON ADVERTISER

No. 10. 60-Point Condensed Woodward, Caps and Lower Case

Terre Haute Chronicle

No. 11. 48-Point Condensed Woodward, Caps

THE HOWARDS REPUBLICAN

No. 12. 48-Point Condensed Woodward, Caps and Lower Case

Weekly Allenton Independent

St. Louis, Mo., U. S. A.

Inland Type Foundry

ELDON STAR

No. 13. 60-Point Edwards, Caps

Rodville Chief

No. 14. 60-Point Edwards, Caps and Lower Case

THE ADAMS SUN

No. 15. 48-Point Edwards, Caps

Doniphan Journal

No. 16. 48-Point Edwards, Caps and Lower Case

INLAND TYPE FOUNDRY

St. Louis, Mo., U. S. A.

334

SMITHTON GLOBE

No. 17. 60-Point Saint Johis. Caps

Daily Sedalia Record

No. 18. 60-Point Saint Johis. Caps and Lower Case

THE LINCOLN GAZETTE

No. 19. 48-Point Saint Johis. Caps

Weekly Pacific Messenger

No. 20. 48-Point Saint Johis. Caps and Lower Case

ST. LOUIS, MO., U. S. A.

INLAND TYPE FOUNDRY

BRASS RULES

The 2-Point Single, Dotted and Hyphen Rules—Nos. 27, 111, 113 and 117—line with all our STANDARD LINE Type Faces with easy justification by means of point system leads and slugs. The following Rules are sold in strips of 2 feet each.

No.	Body	Per foot	No.	Body	Per foot
1	1-Point	$0.05	29	4-Point	$0.16
2	1½-Point	.06	30	5-Point	.20
3	2-Point	.08	31	6-Point	.28
4	3-Point	.12	32	7-Point	.32
5	4-Point	.16	33	8-Point	.35
6	5-Point	.20	34	9-Point	.38
7	6-Point	.28	35	10-Point	.40
8	7-Point	.32	36	12-Point	.50
9	8-Point	.35	37	1-Point	.05
10	9-Point	.38	38	1½-Point	.06
11	10-Point	.40	39	2-Point	.08
12	12-Point	.50	40	3-Point	.12
13	14-Point	.55	41	4-Point	.16
14	16-Point	.58	42	5-Point	.20
15	18-Point	.60	43	6-Point	.28
25	1-Point	.05	100	2-Point	.08
26	1½-Point	.06	101	3-Point	.12
27	2-Point	.08	102	4-Point	.16
28	3-Point	.12	103	5-Point	.20
			104	6-Point	.28

BRASS RULES

The 2-Point Single, Dotted and Hyphen Rules—Nos. 27, 111, 113 and 117—line with all our STANDARD LINE Type Faces with easy justification by means of point system leads and slugs. The following Rules are sold in strips of 2 feet each.

No.	BODY	Per foot	No.	BODY	Per foot
45	1½-POINT	$0.08	70	2-POINT	$0.10
46	2-POINT	.10	72	3-POINT	.14
47	3-POINT	.11	73	4-POINT	.18
48	4-POINT	.18	74	5-POINT	.22
49	5-POINT	.22	75	6-POINT	.28
50	6-POINT	.28	77	8-POINT	.35
52	8-POINT	.35	78	9-POINT	.38
53	9-POINT	.38	79	10-POINT	.40
54	10-POINT	.40	80	12-POINT	.50
55	12-POINT	.50	85	2-POINT	.10
56	2-POINT	.10	86	3-POINT	.14
57	3-POINT	.14	87	4-POINT	.18
58	4-POINT	.18	88	5-POINT	.22
59	5-POINT	.22	89	6-POINT	.28
60	6-POINT	.28	90	8-POINT	.35
62	8-POINT	.35	91	9-POINT	.38
63	9-POINT	.38	92	10-POINT	.40
64	10-POINT	.40			

BRASS RULES

The 2-Point Single, Dotted and Hyphen Rules—Nos. 27, 111, 113 and 117—line with all our STANDARD LINE Type Faces with easy justification by means of point system leads and slugs. The following Rules are sold in strips of 2 feet each.

No.	Body	Per foot	No.	Body	Per foot
110	1½-Point	$0.08	150	1-Point	$0.05
111	2-Point	.10	152	2-Point	.08
112	1½-Point	.08	153	3-Point	.12
113	2-Point	.10	154	4-Point	.16
116	1½-Point	.08	155	5-Point	.22
117	2-Point	.10	156	6-Point	.28
118	3-Point	.14	160	2-Point	.08
120	2-Point	.10	162	3-Point	.12
121	2-Point	.12	163	4-Point	.16
122	2-Point	.12	164	5-Point	.22
123	1½-Point	.10	165	6-Point	.28
124	2-Point	.12			
125	3-Point	.16	200	6-Point	.30
126	4-Point	.20	201	8-Point	.40
130	2-Point	.12	210	6-Point	.30
131	2-Point	.12	211	8-Point	.40
132	2-Point	.12	220	6-Point	.30
133	2-Point	.12	221	8-Point	.40
140	1-Point	.05			
142	2-Point	.08	302	4-Point	.24
143	3-Point	.12	311	4-Point	.24
144	4-Point	.16	321	4-Point	.24
145	5-Point	.22	333	6-Point	.34
146	6-Point	.28			

BRASS RULES

HEAD RULES

Either Double, Parallel or Single. Cut also from any other number of Rule made by us.

No. 49—5-POINT

No. 50—6-POINT

No. 60—6-POINT

No. 75—6-POINT

No. 89—6-POINT

	5-Point	6-Point		5 Point	6-Point
4-Column — 9 inches or less	$0.20	$0.25	7-Column — 16 inches or less	$0.30	$0.40
5-Column — 12 inches or less	.25	.30	8-Column — 18 inches or less	.35	.45
6-Column — 14 inches or less	.30	.35	9-Column — 20 inches or less	.40	.50

COLUMN RULES

6-POINT

7-POINT

8-POINT

	6-Point	7-Point	8-Point		6-Point	7-Point	8-Point
12 inches or less	$0.30	$0.38	$0.40	22 inches or less	$0.52	$0.65	$0.70
16 inches or less	.40	.50	.50	24 inches or less	.55	.65	.75
18 inches or less	.45	.55	.55	26 inches or less	.60	.70	.80
20 inches or less	.47	.60	.65				

CROSS RULES AND DASHES

These Rules and Dashes are the ones most commonly used. Other faces and styles cut to order. Cut to suit any width of news column, usually 13 ems. Send WIDTH of column and LENGTH of face.

Single Dashes
No. 27, face 6 ems, 6c.

No. 39, face 6 ems, 6c.

No. 100, face 6 ems, 6c.

Parallel Dash
No. 48, face 6 ems, 8c.

Double Dashes
No. 57, face 6 ems, 8c.

No. 58, face 6 ems, 8c.

No. 72, face 6 ems, 8c.

No. 73, face 6 ems, 8c.

Wave Dashes
No. 130, face 6 ems, 6c.

No. 131, face 6 ems, 6c.

No. 133, face 6 ems, 6c.

Single Cross Rules
No. 27, 4c.

No. 39, 4c.

No. 100, 4c.

Parallel Cross Rules
No. 48, 6c.

Double Cross Rule
No. 57, 6c.

No. 58, 6c.

No. 72, 6c.

No. 73, 6c.

Wave Cross Rules
No. 130, 5c.

No. 131, 5c.

No. 132, 5c.

BRASS DASHES

These Dashes are most commonly used. Other faces and styles will be cut to order. Cut to suit any width of newspaper column, usually 13 ems. Send width of column and length of face.

Price, Nos. 1 to 20, each, 10c.; Nos. 21 to 56, each, 15c.

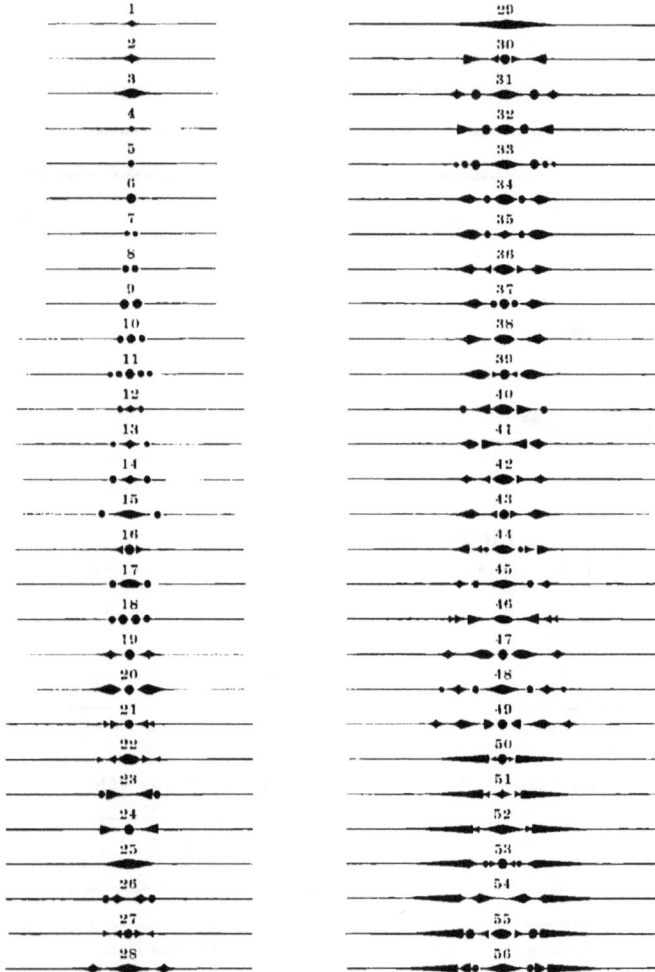

1	29
2	30
3	31
4	32
5	33
6	34
7	35
8	36
9	37
10	38
11	39
12	40
13	41
14	42
15	43
16	44
17	45
18	46
19	47
20	48
21	49
22	50
23	51
24	52
25	53
26	54
27	55
28	56

MONKEY DASHES

No. 27, face 3 ems, 6c.　　　No. 39, face 3 ems, 6c.　　　No. 100, face 3 ems, 6c.

BRASS BRACES

Made in any length. 4-em to 36-em 12-Point kept in stock.

5-em — 12c.

7-em — 12c.

9-em — 12c.

11-em — 15c.

13-em — 15c.

15-em — 15c.

17-em — 18c.

19-em — 18c.

21-em — 25c.

23-em — 25c.

22-em — 25c.

20-em — 25c.

18-em — 18c.

16-em — 15c.

14-em — 15c.

12-em — 15c.

10-em — 12c.

8-em — 12c.

6-em — 12c.

4-em — 12c.

Labor-Saving Brass Rule

The 2-Point Single, Dotted and Hyphen Rules — Nos. 27, 111, 113 and 117 — line with all our STANDARD LINE Type Faces with easy justification by means of point system leads and slugs.
The following Labor-Saving Rules are in stock in 1-pound, 2-pound, 3-pound and 5-pound fonts.

No. 27   2-POINT   Per pound, $1.75

No. 18   4 POINT   Per pound, $1.60

No. 73   4 POINT   Per pound, $1.60

No. 75   6-POINT   Per pound, $1.50

Labor-Saving fonts of any other Rules, excepting Nos. 120 to 133, and 302 to 333, cut to order.

Labor-Saving Brass Rule

The 2-Point Single, Dotted and Hyphen Rules—Nos. 27, 111, 113 and 117—line with all our STANDARD LINE Type Faces with easy justification by means of point system leads and slugs.

The following Labor-Saving Rules are in stock in 1-pound 2-pound, 3-pound and 5-pound fonts.

No. 111   2-POINT   Per pound, $1.75

No. 113   2-POINT   Per pound, $1.75

Labor-Saving fonts of any other Rules, excepting Nos. 120 to 133, and 302 to 333, cut to order.

Brass Ovals

Style D

4

7

10

12

14

18

Styles A, B, C and D kept in stock. See Circles for style. Other shapes and styles will be made to order.

Ovals 1 to 4, each..........$0.80   Ovals 11 to 12, each......$1.00   Ovals 15 to 18, each......$1.50
Ovals 5 to 10, each........ .90   Ovals 13 to 14, each...... 1.25

Labor-Saving Brass Leaders

Standard Line

12-POINT DOTTED BRASS LEADERS   Per pound, $1.40

.... **Labor-saving**
.... Brass Leaders
.... *Are the best and*
.... *Most handy* ....

10-POINT DOTTED BRASS LEADERS   Per pound, $1.50

.... *6, 8, 10 and 12 - Point* ....
.... **Bodies in Stock in**
.... 2, 3 and 5 - pound
.... *Fonts.*

9-POINT DOTTED BRASS LEADERS   Per pound, $1.50

.... *Standard Line Leaders*
.... **line with all our**
.... Romans, Old Styles ....
.... *and Job Faces* ....

....

Special line Labor-Saving Brass Leaders, other than Standard Line, are made to order to match
any face made by other foundries. Send new samples of H and m for line.

Labor-Saving Brass Leaders

Standard Line

8-POINT DOTTED BRASS LEADERS   Per pound, $1.60

...................................Standard Line Leaders..............................
..............................line with our Romans,..........................
.............................Old Styles and Job.......................
.............................Faces.

6-POINT DOTTED BRASS LEADERS   Per pound, $1.60

....These Standard Line.............................
......................................Leaders line with..........................
...................................all our Romans,...........................
....and Job.......................
                                    Faces ....

8-POINT HYPHEN BRASS LEADERS   Per pound, $1.60

- - - - - - - - - - - - -   - - - - - - - - - - - - - - - - -
- - - - - - - - - - - - - - -   - - - - - - - - - - - - - - -
- - - - - - - - - - - - - - - -6, 8, 10 and 12-Point - - - - - - - - - - - - -
- - - - - - - - - - - - - - - - Bodies in Stock in 2, 3 - - - - - - - - - - - - -
- - - - - - - - - - - - - - - - - and 5-pound Fonts - - - - - - - - - - -
-            - - - - - - - - - - - - -
-        - - - - - - - - -
-        - - - - - - - - -
-        - - - - - - - -
-        - - - - - - -

Special line Labor-Saving Brass Leaders, other than Standard Line, are made to order to match
any face made by other foundries. Send new samples of H and m for line.

BRASS CIRCLES

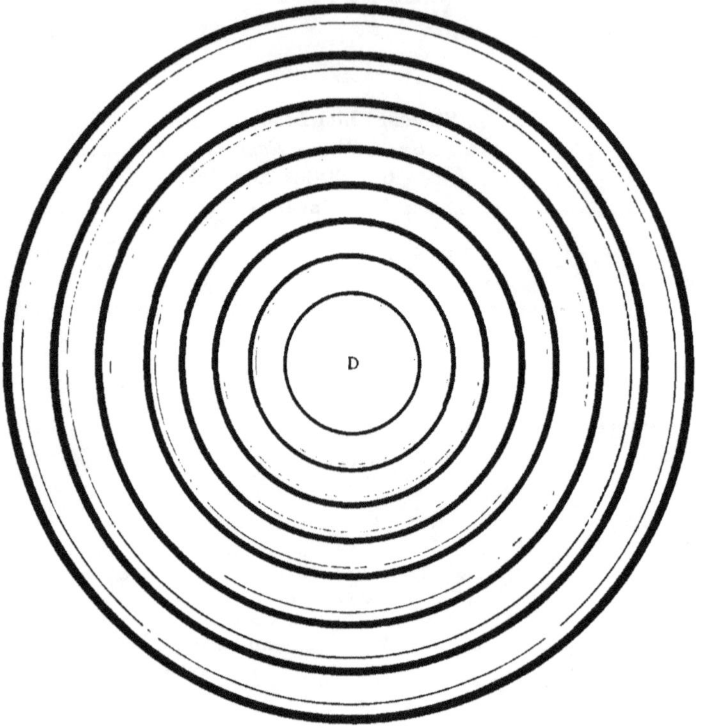

DIAMETERS AND PRICES

Styles A, B, C and D are kept in stock in the following sizes. Other sizes and styles will be furnished to order.

3/4-inch	$0.40
7/8-inch	.40
1 inch	.50
1 1/8-inch	.50
1 1/4 inch	.50
1 3/8-inch	.50
1 1/2-inch	.50
1 5/8 inch	.60
1 3/4 inch	.60
1 7/8 inch	.75
2 inch	.75
2 1/4 inch	.80
2 1/2-inch	1.00
2 3/4 inch	1.00
3 inch	1.00
3 1/4-inch	1.25
3 3/4 inch	1.50

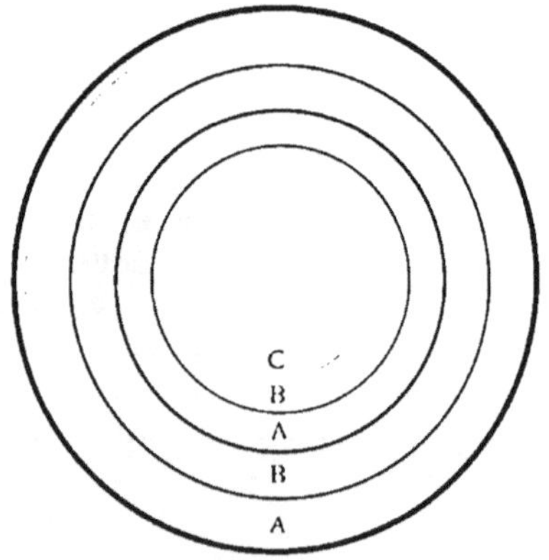

The Circles shown above are made from Solid **SEAMLESS** Hard Brass, and are far more accurate and durable than the ordinary make.

WHAT THEY SAY OF STANDARD LINE

∽

The following testimonials are unsolicited, and we have therefore, to save space, in many cases omitted extraneous matter. The originals, and many other similar letters, are on file in our office.

LONGMONT, COLO., April 4, 1895.
You are on the right track; every foundry in the country will have to come to it.
C. W. BOYNTON.

LANCASTER, PA., July 5, 1895.
Your STANDARD LINE idea is something printers should have had years ago.
D. B. LANDIS.

SAVANNAH, TENN., Nov. 7, 1896.
I am delighted with STANDARD LINE and UNIT SET. See order on another sheet.
C. L. HEFNER.

STAUNTON, VA., Dec. 9, 1896.
I am in need of some new type, and STANDARD LINE, UNIT SET is the thing to save time and make money. P. A. ROSS.

READING, PA., Jan. 5, 1895.
I am glad you have made another step toward perfection. The point system was one; this is even greater. E. D. WESCOTT.

PUEBLO, COLO., April 20, 1896.
We assure you that when we purchase new supplies it will be STANDARD LINE of Inland manufacture. RIVERSIDE PTG. CO.,
per M. D. PENNEBAKER.

BARNESVILLE, O., June 8, 1896.
We like your new series of borders very much, and trust to be able to send you more business, especially in STANDARD LINE type.
HANLON BROS. PAPER CO.

BRADFORD, PA., July 19, 1895.
I am greatly pleased with your STANDARD LINE, and the original ornaments and borders are marvels of beauty. Wishing you success. J. NEVIN HUBER.

GREENUP, KY., Nov. 7, 1895.
I just want to say that during my experience of fourteen years in the printing business I have never worked with material equal to yours. BENJ. POWELL.

DENVER, COLO., Mar. 18, 1896.
I like your system, not only as to its lining feature, but the making of figures for old style type all above the line and of the same height. E. L. WEPF.

ST. ALBANS, VT., Dec. 5, 1896.
I purchased my entire outfit from you when I could have bought for less closer at hand. I like your type very much, and any inquiries from this section refer to me.
C. W. BUCKLEY.

ST. ALBANS, VT., Feb. 10, 1897.
I am entirely satisfied with my type and know there is a big saving in composition.
C. W. BUCKLEY.

BALTIMORE, MD., Mar. 22, 1895.
We are much pleased with your system of lining, and, being more fully in the job line, can perhaps more thoroughly appreciate it than some others.
JOHN S. BRIDGES & CO.

ADRIAN, MICH., Jan. 6, 1897.
I like your STANDARD LINE, having the 8-Point No. 23. STANDARD LINE is great and no mistake; also the UNIT SET, as it saves a great deal of time.
F. W. NICHOLS.

MINNEAPOLIS, MINN., April 20, 1896.
We are very much pleased with the type and highly elated that we chose it in preference to the "Oliphant" and "Livermore."
UNIVERSITY PRESS OF MINNESOTA,
T. H. COLWELL.

PLESSIS, N. Y., Dec. 26, 1896.
I admire your STANDARD LINE. I have some of it, bought from Golding & Co., and must say I like it exceedingly well. What a lot of bother it saves in my work.
U. E. BROWN.

RIDGEWAY, ILL., July 6, 1895.
The bill of type we ordered on the 20th ult. was received in good shape, and to say we are stuck on your material does not express it. Accept thanks for promptness.
CURRY & BLAIR.

PAWTUCKET, R. I., Feb. 28, 1895.
Your system of type-making is certainly a great improvement over the common way, and we have no doubt but that the type will pay for itself in the amount of time saved.
H. H. BEVIS.

WHAT THEY SAY OF STANDARD LINE

CARTERVILLE, ILL., Jan. 6, 1897.
Enclosed find check for $15.00. Please credit my account for that amount. Also find order for some of your borders, which please ship by express at your earliest convenience. The more I use STANDARD LINE type the better I like it. C. BUSH.

HARTFORD, CONN., Mar. 8, 1895.
I think your style of making type to line an admirable innovation in the type line. This, together with the point system make two of the greatest boons that I have known in my thirty years' experience in the printing business. I wish you success in your new adventure.
W. H. BARNARD.

MACUNGIE, PA., Aug. 13, 1896.
We feel determined to get some, if not all, of our type eventually of your excellent productions. We have some of them now, and, as you assert, your STANDARD LINE is the best thing out for the practical printer who is working not alone for "glory." I wish you success. O. P. KNAUSS.

CARUTHERSVILLE, MO., Sept. 7, 1896.
STANDARD LINE has been in use in the office now for several months, and to say that all concerned are more than pleased is putting it mildly. Better results are obtained with it in both ordinary and color printing, and our printers say they like it better than any they have used. LONGGREAR PTG. & PUB. CO.,
per DEL LONGGREAR.

NASHVILLE, TENN., Mar. 19, 1896.
Our foreman has made a test of the value of your type as compared with other makes and says: "The value of your STANDARD LINE type can only be appreciated by using it." Your scheme is a good one, and will save many dollars in time consumed in cutting leads, cardboard, etc., to force justification. The far-seeing master printer will avail himself of the STANDARD LINE type without hesitation. We hope to favor you with an order very soon. MARSHALL & BRUCE CO.

CARTERVILLE, ILL., Nov. 6, 1896.
As McKinley has been successful, we feel that there is sure to be a business boom and believe that it is our duty to prepare for what is sure to come. You may rest assured in our purchases the order will call for STANDARD LINE type—we will use no other. Enclosed

find order for series of your Saint John—to our mind the best series for general job work turned out by any foundry this year.
COPELAND & BUSH.

MINNEAPOLIS, MINN., Sept. 30, 1895.
We have had some of your type and shall soon want more. I congratulate you on your splendid improvements and wish for you that substantial support which your enterprise so richly merits. I am very much interested in the developments you are making to place the system of type where it should have been long ago.
"THE NORTHWESTERN MILLER,"
per HENRY HAHN, foreman.

MINNEAPOLIS, MINN., Feb. 24, 1896.
Your STANDARD LINE type has given us the best of satisfaction. It is all you claim for it. The most difficult composition and complex justification can be done accurately and economically. You are so far ahead in the race for supremacy that it will require the utmost effort on the part of your competitors if they would succeed in closing up the gap.
"THE NORTHWESTERN MILLER,"
per HENRY HAHN, foreman.
[P. S. — Since writing this letter Mr. Hahn has gone into business for himself, purchasing his entire outfit from us.]

MINNEAPOLIS, MINN., Nov. 12, 1896.
I did not order a line of type from any other foundry. Mr.........., the resident manager of ...., was quite exercised to think that I could put in so good an office as I did and get along with only your type. I assured him that I not only could do so, but that I could turn out the very best kind of work with your STANDARD LINE type. It worried him not a little, and I think he had good cause, judging from the expressions of the many printers who have visited the office, many of whom know little of the many advantages of your STANDARD LINE. They are beginning to appreciate it now, and I am confident that it will be to your advantage to have such an office in the city. Mr. Harmon heard of my starting in and we corresponded, with result of the firm of Hahn & Harmon. We are very much pleased with the type and we hope in the future to do considerable business with you. HENRY HAHN,
of HAHN & HARMON.

WHAT THEY SAY OF STANDARD LINE

WACO, TEX., Sept. 14, 1896.

No practical man can fail to see the advantage of your system. It is a wonderful time-saver, and time is money. I attempted to make a combination line recently from 10 and 12-Point type from another foundry, and had to use cardboard and paper. 'Tis needless to say I have not tried it again. In matters typographic you have the world by the tail and a down-hill pull. GEO. C. MARTIN.

GLENS FALLS, N. Y., Mar. 14, 1895.

One of the most important advantages of your type, and the one which will no doubt make your foundry rank among the first in the land, is the STANDARD LINE. It has often been a matter of wonder to me that this idea of lining was not given more consideration when the point system was introduced. Any job printer who has spent many weary hours cutting cardboard, paper of all weights, etc., with a little swear word here and there, to line type or different sizes and faces with each other, will fully appreciate your innovation, which will relieve him of one of his greatest composing-room troubles.

JOHN CHAMBERS.

DANVILLE, VA., May 7, 1896.

I have from time to time been the means of securing patronage for you from my present house, and have been hopeful of sending you at some time a better order than heretofore. I am now able to do so. I expect to make a change, and take an interest in the business of E. R. Waddill, and as the selection of some additional type, etc., to be added to his present plant fell in my choice, I decided on yours, believing your system of lining the best in existence. The order for this will be mailed to you to-night, and I hope you will execute and ship at once. I merely write this to show my appreciation of your efforts in giving the printers of the country something they should appreciate. Wishing you success with the new system. J. B. THORNTON.

DECATUR, ILL., Dec. 8, 1896.

The more we work with STANDARD LINE type the more convinced we are that it is a big improvement over the old style "stuff." Figuratively speaking, we have been kicking ourselves for some time because we did not put in our entire outfit out of your STANDARD LINE. We are so well convinced of its superiority that if we were going to buy a new outfit to-morrow we would not have a single face in the outfit that was not STANDARD LINE. Whatever additions we make, in fact what we have made since we really became acquainted with the merits of your type, we have decided will be the best type made — STANDARD LINE. As fast as material is discarded we will put in your type also. This is not in the least flattery, but actual fact from knowledge gained by experience. PENNINGTON BROS.

HARRISBURG, PA., Aug. 15, 1896.

We have now handled the STANDARD LINE type received from you sufficiently to be able to make some conclusions in regard to it. There is unquestionably a desirable economy in justification, so that the same compositor can set more of this type in a day than of the ordinary type. The set of your type bodies would seem to be trifling, and yet our compositors, who work on the new 10-Point by the piece, find they can set more type in a day, and the writer who is somewhat of a crank on accurate justification, has much more satisfaction in passing his fingers along lines down the side of the galley, for it seems almost impossible for a compositor to avoid clean, uniform justification.

J. HORACE McFARLAND CO.,
by J. HORACE McFARLAND.

HYDE PARK, MASS., May 2, 1896.

Yur various shipments of type wer duly and promptly reselved. On the same Monday afternoon I put into a Boston type foundry, that is eight miles from my door, an order for sum sorts. At the same time I placed in the Boston post office three letters, two to different firms in New York for special sorts that I wanted and the letter I sent yu. On the folloing Saturday morning I herd from one New York foundry. I thought that was wel. But in the afternoon of the same day the expresman deliverd your pakage. That was better yet. The Boston firm has not yet sent the sorts I wanted, altho they repeatedly promist to cast them in a few days. I hav spent a number of years in the west and can appreciate western rustling.

A. L. LODER.

[Mr. Loder iz a hearti advocate ov speling reform, hwich accounts for the absens of unnesesari letters in hiz communicashun.]

INDEX TO SPECIMEN BOOK

INDEX TO SPECIMEN BOOK—Continued

INDEX TO SPECIMEN BOOK—Concluded

www.ingramcontent.com/pod-product-compliance
Lightning Source LLC
Chambersburg PA
CBHW021527210326
41599CB00012B/1415